Charging for Irrigation Services
Guidelines for Practitioners

Frank van Steenbergen
Gez Cornish
Chris Perry

Charging for Irrigation Services Guidelines for Practitioners

Frank van Steenbergen
Gez Cornish
Chris Perry

HR Wallingford
Working with water

META
META Research

IWA
Publishing

Published by IWA Publishing, Alliance House, 12 Caxton Street, London SW1H 0QS, UK
Telephone: +44 (0) 20 7654 5500; Fax: +44 (0) 20 7654 5555; Email: publications@iwap.co.uk
Web: www.iwapublishing.com

First published 2007

British Library Cataloguing in Publication Data
A CIP catalogue record for this book is available from the British Library

Library of Congress Cataloging- in-Publication Data
A catalog record for this book is available from the Library of Congress: Control Number: 2006937672

ISBN: 1-84339-141-4
ISBN13: 978-1-84339-141-8

Executive Summary

Charging for Irrigation Services: Guidelines for Practitioners

Frank van Steenbergen
Gez Cornish
Chris J Perry

Irrigation is the dominant consumer of fresh water world-wide, accounting for as much as 80% of use in many water-short countries. Two issues dominate the problems in water resources management generally, and especially the management of irrigation systems: *scarcity of water* to meet competing demands, and *scarcity of funds* to finance operation, maintenance and renewal of existing facilities.

Various international conferences, donor policies and academic papers have pointed to the contribution that appropriate irrigation service charging systems can make to both problems. Higher charges should mean lower demand and/or allocation of scarce water to more productive uses, and higher charges automatically generate revenues to pay for provision of the service.
In consequence, over the last decade, much emphasis has been placed on the need for appropriate charging policies. Different approaches have been advocated, including, the treatment of water as an economic good, exploring the role of demand management, use of full cost pricing, or resource pricing, the introduction of volumetric charging, water markets, or tradable water rights. While the usefulness of at least some of these ideas has been demonstrated in the municipal and industrial water supply and sanitation sectors, their relevance to irrigation was less certain -- though theoretically attractive.

In this context, the UK's Department for International Development (DFID) under its Knowledge and Research Program commissioned a series of studies to review the issue of charging for irrigation services, comprising:

- A detailed literature review, undertaken to assess the lessons learnt from existing experience around the world and to make full use of existing thinking on the subject and experience in almost 50 countries.
- Six case studies of irrigation charging systems in five countries, India (two case studies), Macedonia, Morocco, Nepal and Pakistan, to assess current practice and performance under a range of technical, socio-economic and organisational conditions. The case studies were completed in mid-2003. The results of these two components were synthesized in Cornish, Bosworth, Perry and Burke (2004)..
- The preparation of detailed Guidelines for those responsible for the design and implementation of effective charging policies designed to meet specific local objectives. A summary of these first two components is included with the Guidelines.

The literature review confirmed that the rationale and theoretical background for a successful system of irrigation service charges (ISC) is complex. The literature also shows, as confirmed by the case studies, that solutions must be location and situation-specific. In consequence, the full Guidelines are lengthy. For policy makers and others not directly concerned with the details of implementing an irrigation charging system, this summary provides essential guidance to the basic issues which may be supplemented by reference to specific parts of the Guidelines. The summary is designed to highlight an issue that many of the recent recommendations have failed to recognise, namely: A successful system of irrigation service charges comprises a framework of interdependent definitions, decisions, actions and physical infrastructure. Any intervention or initiative must recognise these interdependencies, and failure to do so will inevitably result in an unsuccessful system and discredit the objectives that irrigation service charges legitimately serve.

The literature review and case studies confirmed that there are two common and related flaws in many attempts to introduce or improve ISC: first, a failure to recognise the individual importance of the two key elements - the *service* and the *charge.* Many irrigation systems have a poorly specified level of service, and the scope for charging is in consequence limited. Attention to how well and clearly the level of service can be defined will clarify the appropriate basis for charging. Second, many initiatives to introduce ISC are imposed by external agencies with insufficient attention given to the local context. Objectives are predefined, the costs that should be recovered are specified, who should pay and the nature of payments are dictated, as is the format of the charging system. Too often these recommendations are inconsistent with existing laws, unsuited to the countries' objectives, and possibly even impossible to implement with available infrastructure. In general, for the last ten years there has often been an unrealistic expectation that irrigation water pricing in one way or another would make a major contribution to demand management and would cause significant improvement in farm level water efficiency. There little practical evidence to substantiate this effect.. There are several explanations:

- The fact that in many irrigation systems the challenge is not so much the pricing of water but the effective collection of irrigation charges;

- The fact that in many irrigation systems water is a vital, but low priced input (compared to other production costs) - the price elasticity is therefore low. Moreover very significant changes in the price of public irrigation services are in most countries politically infeasible:

- The fact that in many systems individual farmers water use is conditioned by the way water is delivered from the main system. Individual farmers response to prices are less relevant.

These Guidelines seek to avoid the shortcomings of a preconceived strategy and does not intend to give a prescriptive approach. Rather the reader is taken through a logical sequence of steps that start from the present situation, identify the country's objectives for ISC, compare alternative approaches to achieving the objective, identify the most effective way to meet specified objectives, and identify the constraints that must be addressed (through revised policies, rules, or infrastructure) to implement the selected ISC system. The Guidelines moreover argue to put irrigation water charging in the context of the overall financial management of the irrigation systems. This requires one to equally pay attention to rationalizing expenditures and to develop other source of revenue than merely depending on irrigation charges: ranging from charges to large domestic water users to developing waterfront property.

Structure of the Guidelines
The structure of the Guidelines reflects the interdependent framework of issues that must be addressed in formulating a successful ISC system. The steps involved relate to:

- The Policy Environment - What laws and formal statements of policy by relevant authorities and other government ministries, govern the imposition of ISC?
- Setting Objectives - What is the primary objective of ISC? What additional objectives are of interest?
- The Basis for Charging - What will be the basis for computing the ISC - water delivered, area cropped, crop type?
- Quantifying the Charge - How is the amount to be recovered calculated?
- Implementation - Who will collect ISCs? Who will pay? When? What sanctions will apply for non-payment?

Annex A of the guidelines provides a series of checklist tables that lead the user through the logical steps set out in the body of the text. It is recommended that those checklists provide a working basis when applying these guidelines to design or revise an ISC system.

This framework is in part hierarchical - the Policy Environment will either *define* the objectives, rules and procedures for setting irrigation service charges, or provide *guidance* on the principles to be followed. Often there will be gaps in the policy environment so that clarification of key issues is a precondition to any further progress. Sometimes the process will be iterative - for example, if charges based on volume of water delivered are planned, the infrastructure must be reviewed and perhaps upgraded to ensure that accurate measurement at the desired points in the system is feasible.

The outcome of the entire ISC process must be a system that meets defined objectives while being internally compatible with governing legislation, and technically feasible. The process must therefore consider political, legal, administrative, technical and operational aspects – as well as ensuring acceptance by the stakeholders.

Definitions
The terms charge, price, cost and value are commonly used interchangeably. To avoid confusion, the terminology used in the rest of this summary, and throughout the Guidelines is consistently based on the following definitions:

- *Irrigation Service Charge*: the total payment made by a user for an irrigation service. It may comprise fixed elements (e.g. US$ 20/ha) plus variable elements (e.g. US$ 1 per thousand cubic metres of water). In this example, if a user with one hectare took 10,000m^3 under the above charging system, the charge would be US$ 30.
- *Price*: in the above example, the average price of water would be the total charge divided by the total quantity of water received (US$ 30 / 10,000 = US$ 0.03/m^3). The marginal price would be the cost of an additional unit of water ($ 1 / 1000m^3 = US$ 0.0001/m^3)
- *Cost of the irrigation service*: the expenses incurred by the supplying agency in providing the service. Precise definitions depend on local rules, but typically include operation, maintenance, staff and fuel costs, plus some elements of replacement costs and amortisation of capital.
- *Value of water:* incremental income received by the farmer as a result of irrigation services, divided by the quantity of irrigation water used.

Policy Environment
Clear policies assist the formulation of ISC, and policy support is required for any chosen ISC to be viable. The existing irrigation service provides the point of departure for identifying elements of policy that are essential to the definition of the service:

- What powers do the State and other authorities have in developing, managing and allocating water?
- How much water is allocated to irrigation?
- What priority does irrigation have in case of shortage in other sectors?
- How is the user's water allocation determined, specified and measured?

The answer to these questions will define the nature of the service provided. They also begin to identify the roles and responsibilities of the various participants in the system. Review of the existing system of charging will also assist in defining:

- Who owns the irrigation facilities?
- Who is responsible for maintenance, operation and replacement of facilities?
- What policies are specified for recovery of costs incurred by government agencies and which costs are to be recovered (operation, maintenance, replacement, investment)?
- What is the policy for cost-sharing where activities (e.g. dam safety programmes) or infrastructure serve multiple beneficiaries?
- Who is responsible for assessing and collecting the irrigation charge?
- Does present practice work?
- Who receives the collected charges and for what purpose are they used?

- Do circumstances exist where charges are waived or reduced in order to avoid imposition of financial hardship on users? Does the system work or is it abused?

The policies listed above may be derived from legislation, as well as stated government policies, in a variety of sectors other than irrigation (e.g. water supply, agriculture, environment). Policies may also specify, on socio-economic grounds, situations where charges are to be set on grounds other than financial sustainability.

Where the specified policies of the government (e.g. full cost recovery) are not achieved, it is important to carry out an analysis of current practice to identify why policy is not being fulfilled. For example, is the policy unrealistic, is the charging system compatible with the primary objective, do the users have a legitimate reason for non-payment of charges, are the agencies responsible for assessment and collection significantly under-resourced or do they lack incentives to implement unpopular polices?

Where there are gaps in the policy, the ISC system will need to specify, or presume, a policy, which in turn will require political and legal endorsement. Closing these gaps, with political commitments (and where necessary, legislation) is a critical step in the formulation of successful ISC. If new proposals for ISC generate conflict with existing policies, it is essential to revisit the policy environment for endorsement and legislative backing.

Setting Objectives
The three most widely stated objectives of irrigation charging are:

- To achieve a specified and consistent level of cost recovery from users.
- To provide an incentive to irrigators to reduce water consumption (demand management).
- To increase the productivity of water at the individual user level, or through transfers to more productive users or uses, with the objective of increasing the level of economic benefit per unit of water.

The detailed interpretation and realisation of these objectives is not simple and the charging systems used to achieve each may actually conflict, so it is important to consider carefully the primary objective of the proposed ISC system.

"Cost recovery" can range from simple day-to-day operational costs to the entire cost of operation, maintenance, future replacement and amortisation of past investments. Where a degree of capital cost recovery is required, the term of recovery and interest rates to be charged must also be defined. Government policy should be set out which costs are to be recovered. The selected ISC must be consistent with that policy and supported by legislation.

Incentives can achieve a range of "demand management" objectives:

- Any volume related charge for irrigation services provides an incentive to avoid absolute wastage of water such as letting irrigation water run to drains.
- Higher charges - if suitably designed - can make some uses unprofitable, hence reducing demand.
- Charges can theoretically be set at a level that limits demand for water to some desired level, a so called "market clearing" price.

Setting the ISC so as to achieve cost recovery, once the level and components to be charged have been determined as a policy issue, is relatively straightforward. Any system of ISC will achieve cost recovery if the charge is set at the appropriate level. This is not to suggest that implementation and collection are necessarily straightforward - there may be many problems and controversies - but defining the required ISC is simple.

Demand management objectives are more difficult to realise and several of the limitations are described earlier in this summary. It is always instructive to estimate the value of water in relation to the likely price. For example, if a farmer's net income with irrigation is US$1,000 per hectare higher than without irrigation and the quantity of water used is 10,000 m^3/ha, then the productivity of water is in the order of 10¢/m^3. If the ISC to meet cost recovery objectives is US$100 / ha, then it is clear that a volumetric charge sufficient to meet cost recovery objectives (i.e. 1¢/m^3) will do little to limit demand. A volumetric charge will not serve the purpose of demand management to the point of balancing supply and demand but may encourage a degree of care in water use.

An additional complexity in respect of demand management is the important distinction between water *diverted* to irrigation use, and water *consumed*. Irrigation efficiency, often computed as the ratio of diversion to consumption, is usually well below 50% in surface systems. But the "lost" water is sometimes still available for use, being either captured in drains and returned to the river or recharging aquifers for use through wells. Thus, increasing the efficiency of irrigation through charging mechanisms may not save much water. In the extreme case, when water is surplus in one season and scarce in the next season, "inefficient" irrigation to recharge aquifers may be positively beneficial. However, the reaction of a farmer to higher volumetric water charges may be to improve his technology so that *more* of the water he receives and pays for will be consumed. This in turn generally leads to higher water productivity, because more of that water is converted productively into crops. So, paradoxically, the result of an increase in volumetric price, inducing a shift in technology, is to increase the demand for water as long as land is scarce.

Notes:
1. Revenue level and stability depends on selected charging basis.
2. Ease of administration for quotas depends on the nature of quota: proportional division of water, uniformly across the irrigated area provides an exceptionally simple system to administer; seasonally variable allocations with varying schedules of delivery are complex to administer.
3. Tradable water rights are complex to administer, combining the difficulties of a sophisticated quota system with the additional need to adjudicate on the third-party impacts of transfers.

The final consideration from the agency's perspective is whether the irrigation service implied by the charging basis is technically feasible. Area and crop-based systems do not necessarily specify the nature of the irrigation service as an input (volume, flow rate, timing, duration) – rather the service is defined on the basis of the outcome (a certain area to receive water and a crop matured successfully). Volumetric charging, on the other hand, requires detailed measurement of the water supply. Furthermore, if there is to be an incentive to use less water at the individual farm level, the possibility must be in place to provide a controlled and individualised service at that level. Tradable water rights also require this differentiation of supply at the level that water is to be traded. Very few surface irrigation systems in developing countries are capable of such operation. Only a proportionally based quota system such as *warabandi* avoids the need for adjustments to deliveries at the individual farm level, making management much easier.

Quantifying the Charge

Earlier sections have noted the need for a clear policy regarding the scope of the costs to be recovered. It is assumed here that this step has been completed and the issue to be addressed here is primarily the procedure for accounting and cost allocation, and the means of transparently reporting this information. It is further assumed that one objective of the charge is to recover a specified level of the costs associated with service provision – establishing a clear linkage between cost of service and payment for service. Where this is the case, it is likely that levels of charge will vary between schemes, or possibly within a single large scheme if water is drawn from different sources with different attendant costs.

The level of charge, together with any agreed subsidy from government, must cover the costs of operating, maintaining and replacing the system – otherwise the system will deteriorate and the vicious circle of inadequate funding/deterioration/poor service/unwillingness to pay will start. If it is intended to reform an existing charging system where payment is not linked to cost, then a well-defined and transparent transition period must be built into the reform timetable with either the level of subsidy or the shortfall in income phased out as charges rise.

If a specified level of cost recovery is to be achieved, the computation of the ISC must be based on measured, actual expenditures, appropriately allocated. Many existing irrigation departments do not keep such records. The department responsible for irrigation services may also be responsible for other activities such as drainage, flood control, domestic water supply, generation of hydropower, etc. They may not keep accounts in a way that allows identification of those activities that fall under the agreed definition of recoverable irrigation costs. Clear agreement must be reached with all relevant stakeholders on how costs will be allocated among beneficiaries, for example, where channels serve a number of "irrigation groups" plus a village water supply. In such cases, it is essential that rules specifying how costs are divided between different users are agreed, understood and applied.

Reorganising the accounts to meet the needs of cost recovery is a challenging task, which has two-way benefits. First, it clarifies to the beneficiaries, and the population more generally, what expenses are incurred in providing the service, which helps build consensus that payment is appropriate. Second, it clarifies how much the providing agency is spending to provide the service, which can help identify areas where economies can be made by rationalising expenditures: reducing expenditures on major cost items, such as energy bills, or closing down non productive sections of the irrigation systems. In general, a 'business plan' approach to irrigation financing is advocated – whereby opportunities for generating income from other sources than irrigation charges are explored and developed.

Particularly where inflation is severe, formal procedures for updating costs retroactively are essential – in any event a regular procedure for updating charges must be an intrinsic and automatic part of the process. Most importantly, where cost recovery is to include costs for replacement, a procedure is needed to recover on average more than the "typical" O&M costs and set this sum aside for the time when major replacements are needed.

It may be effective to establish a panel of experts or a committee to oversee aspects of this process. The benefit of such an approach is to neutralise the political dimensions of the debate - but any committee must have clear terms of reference that prevent the recommended charges being used as a political lever.

Implementation

The previous sections have dealt with the considerations involved in designing and formulating a viable charging system. These are necessary, but not sufficient, conditions for a successful system.

Strategies for water charging depend critically on the effective assessment and collection of water charges. An efficient charging system will achieve a high revenue performance with low transaction costs. If charges are not seen to be imposed and collected fairly, then payments from all beneficiaries are likely to fall, and revenues to the operating agency will be insufficient to meet its costs, triggering a vicious circle of decline. If the process followed has been based on existing rules, or new rules modified through political consensus and with legal backing, then the commitment of politicians and the law should be in place.

Depending on the selected basis for charging, implementation will involve measuring and recording areas or water delivered - and the infrastructure (physical and organizational) must be in place to ensure that the records produced are acceptable to both the providing agency and the beneficiary. To avoid systematic errors (whether deliberate or accidental) cross-checking of a proportion of records should be considered - at least in the early seasons of introducing a new system. Transparency builds trust, and the records of service delivered and payments made should be open to inspection, as should the accounts of receipts and expenditures by the operating agency. In many cases low payment is not so much caused by a limited capacity to pay but by a lack of confidence in the fairness of the services and charges, in other words a low willingness to pay. Intriguingly it is often irrigation systems where charges per hectar or per unit of water are low that suffer from low collection records. In many cases such low charges in fact result in a low 'willingness to collect'. Procedures for enforcement in cases of failure to pay should be defined and made public and require backing in law.

Acknowledgements

In addition to the work of the principal authors of these Guidelines, important contributions were made during the course of their formulation by CMS Consultants (Nepal), Alan Hall, Chris Finney, Nawal El Haouaru, John Skutsch. Several reviewers also made constructive comments made on earlier drafts: Ramesh Bhatia (India), Hartmut Bruhl (Germany), Simi Kamal (Pakistan), Waltina Scheumann (Germany) and Loh Kwai Sim (Malaysia).

Contents

Contents continued

Tables

Figures

Contents continued

Boxes

Annexes

1. Background to the Guidelines

1.1 INTENDED AUDIENCE

The Guidelines address conditions occurring predominantly in less developed countries, typically including small farms, poor farmers, low levels of charge and recovery, inadequate funding of operation and maintenance, and high demand for water. They are targeted at water managers[1], who may work at national or regional level who may be considering the introduction or reform of irrigation charging, and staff working at scheme level who must implement such policies. They should also be of interest to donor agencies and consultants who work with government water agencies and Water User Associations. The Guidelines may also be of use to NGO and civil society groups interested in sustainable management of irrigation services. They are aimed at practitioners rather than being a contribution to academic debate. They are based on relevant economic theory and practical field experience in a number of developing countries but it is not the intention to provide novel theoretical arguments related to pricing. However, it is useful for the reader to have some understanding of concepts and for that reason Annex B provides a simple summary of the basic economic theory.

> ***The Guidelines are based on relevant economic theory and practical field experience in a number of developing countries.***
>
> ***They identify the objectives of irrigation service charges; the impacts of charging systems on water users and water supply agencies; alternative means of balancing supply and demand for water; and the actions necessary to apply different charging systems.***

1.2 STRUCTURE AND CONTENT OF THE GUIDELINES

The content of these Guidelines is based on the findings of a literature review and case studies carried out under this project (see Section 1.5). In view of the continuing focus on economic instruments as a possible means to address the scarcity of both water and funds within the irrigation sector, these Guidelines aim to set out the various *objectives* of irrigation service charges; the *impacts* of different charging systems on water users and water supply agencies. Objectives fall into two categories: first, to increase the revenues from the users of water services so that they are financially self-sufficient, and second (where water is scarce) one or more of a variety of "demand management" objectives. These include *reducing consumption* of water (for example by restricting unrecoverable losses), *improving the productivity* of water (for example by growing higher value crops or transferring water from irrigation to higher value uses), and ultimately *balancing supply and demand.* Finally, the *preconditions* and *actions* necessary to apply different charging systems are considered.

The Guidelines are primarily concerned with public sector, surface irrigation. Private irrigation charging systems and levels of charge are normally a matter for the individuals and entities involved, without direct government intervention. The frequent success of private development in achieving sustainable financial management over many years serves to emphasise that financial sustainability is feasible, provided an adequate institutional framework is in place and accepted by the stakeholders. No special attention is given to groundwater irrigation. Although there may be differences in detail from surface water, the issues of financial and resource sustainability are common. However, the greater part of groundwater irrigation is in the private sector.

Separate charging for drainage on public irrigation schemes is not covered in the Guidelines. In most developing countries, drainage is included in the irrigation charge rather than being charged for separately. Otherwise, where drainage boundaries and irrigation boundaries are different, the systems for charging for these services are also separate. Furthermore, drainage charging provides little or no scope for saving water.

1.3 CONTEXT

Irrigation agencies commonly face two major challenges: increasing competition for the resource itself - *scarcity of water*; and inadequate funding to meet the needs of current O&M and future investment - *scarcity of funds*.

Scarcity of water manifests itself in various ways:

---

[1] The Guidelines are aimed at a broad audience – from consultants, through operational staff, to policy and decision-makers. We use the term "manager" generically to reflect a decision-maker in any of these roles.

- In some irrigation systems, "head end" irrigators take whatever water they want, while users at the tail go without in times of shortage.
- Unsustainable pumping from aquifers (pumping in excess of the average rate of recharge) provides a temporary source of water, sometimes for years, but ultimately the source will fail and developments dependent on that source will collapse.
- Rivers that used to flow year round, maintaining estuarine ecologies and flushing waste to the sea, are in some cases now dry for months in every year.
- Non-agricultural demands are competing with existing irrigation uses, and pollution from industries and domestic users sometimes renders drainage flows unfit for recycling into irrigation.

Scarcity of funds for operation and maintenance (O&M), rehabilitation, and capital improvements is also a widespread problem in publicly funded irrigation schemes. While the O&M costs of irrigation schemes vary widely a number of features are common to most schemes in developing countries. First, many governments have been wary of increasing or even enforcing charges to users, who are often poor farmers. Second, the staffing of government agencies has generally expanded, with the objective of creating jobs. Third, the emphasis of governments has been on promoting economic development, sometimes neglecting fiscal prudence. In combination, these factors have meant that revenues from irrigation charges have failed to keep pace with the costs of O&M even on schemes where these costs are relatively low.

> ***In theory, where water resources are over-committed, economic demand management tools allow policy makers to avoid difficult allocation decisions – allowing market forces to perform these unpopular tasks.***

Water scarcity has conventionally been managed by means of legal and regulatory measures setting priorities and allocations, with formalised procedures for resolving disputes. More recently, increasing attention has been given to the use of economic instruments- generally interpreted to include pricing and water markets - as a means to address the problems of excess demand for water and deficiency of funding simultaneously. Charging the full cost for irrigation services will generate revenues needed for financial sustainability, which will help towards improved management of the resource. It is also often anticipated that by raising charges demand may fall, thus saving water. The use of economic instruments for demand management is seen as having a variety of additional benefits, such as: reduced pressure to construct new facilities to augment supply, encouraging economies in water use, reallocation from less productive to more productive uses, and allocation based on decisions of users rather than "top down" command and control mechanisms. Additionally, in cases where water resources are already over-committed, economic demand management approaches, in theory at least, allow policymakers to avoid difficult allocation decisions- allowing "the market" to perform these unpopular tasks. However, in practice even the recovery of basic operation and maintenance costs remains elusive.

1.4 DEFINITION OF TERMS

It is important to define the basic terms and concepts used throughout these Guidelines at the very outset and such terms are listed below. Definition of the various charging systems available to managers is provided in chapter 4 where they are linked to the various objectives of irrigation charging.

"Irrigation service charging" (ISC) means any system through which a charge is levied on the beneficiary of irrigation services - ranging from a land tax, whose rate is affected by the availability of irrigation services, to direct volumetric charges for irrigation water.

The terms prices, charges and costs are commonly used interchangeably – the *price* of apples is what they *cost*; the *price* is what a seller *charges* a buyer. But in this document we try to maintain a clear distinction between concepts of charge (the total payment made, with units such as dollars) and price, which is always "per unit of water", and thus has units of $\$/m^3$.

The *charge* is the total payment made by a user for an irrigation service. It may comprise fixed elements, for example, US$ 20/ha whether irrigation is used or not, plus variable elements, for example, US$ 1 per thousand cubic metres of water. In this example, if the user with one hectare took 10,000m^3 under the above charging system, the *charge* would be $30, the *average price* (total charge/total quantity of water received) of water would be $0.03/m^3, and the *marginal* (incremental charge per additional unit of water) *price* of water would be $0.001/m^3.

Where the charge is based on something other than the actual volume of water delivered - farm area, area cropped, etc. - the *marginal* price is usually zero: if the farmer gets a more, or less, water the charge remains the same.

The total *charge* and its component parts are the variables to which users react when deciding whether to irrigate, what crops to irrigate, and how much water to apply.

The *cost* of the irrigation service comprises those expenses that the supplying agency incurs in providing the service. This may include staff costs, operational costs, maintenance costs, costs of replacement and modernisation, and amortisation of capital costs.

The *value* of water has been computed in the case studies as the return to irrigation water after deduction of identifiable input and labour costs from the gross value of production. As defined, this is only an indicative figure, ignoring rental values of land, depreciation of assets, taxes and subsidies on inputs and outputs. This approach extends the analysis adopted by the International Water Management Institute in assessing returns to irrigation - moving towards a net return to irrigation water, rather than the gross return adopted in IWMI's work (Molden, 1997).

Other bases for estimating the value of water are found in the theoretical literature, as well as broader definitions of the *resource cost of water*, which encompass externalities such as pollution impacts, inter-generational distribution of benefits, environmental impacts, and so on. These concepts are extremely difficult to measure or quantify and are highly sensitive to underlying assumptions - for example the different 'value' of incremental income to poor versus wealthy recipients. As a consequence, such factors play a limited role in the practicalities of setting irrigation service charges. Only in circumstances where a clear relationship can be established between an activity and its associated costs is it normally possible to incorporate such costs directly into the general scheme of ISCs. For example, the costs of treating effluent discharged into a river.

1.5 LESSONS FROM THE LITERATURE[2] AND CASE STUDIES[3]

A literature review, carried out under this project, provided some indication of key general issues related to water pricing:

- Cost recovery is the dominant objective of irrigation service charges; demand management is mentioned occasionally.
- Socio-economic concerns of resource allocation between sectors, pollution charging and benefit taxation are recorded in the theoretical literature but are seldom the key drivers of national policies concerning water pricing.
- Cost recovery and water demand management are separate objectives, which require different types of intervention. It is common to find substantial documents where these different objectives are apparently interchanged at random. Clarity of objectives is essential so that policy makers have a clear understanding of what they are seeking to achieve and the tools that are relevant to that purpose.
- Where the objective is cost recovery, most programmes seek only to recover annual O&M costs and possibly some fraction of capital investment costs.
- Cost recovery can be realised without volumetric water pricing, thus minimising technical and administrative complexity.

---

2 Summarised from Bosworth et al. 2002

3 Summarised from Cornish and Perry, 2003

The volumetric water prices required to achieve cost recovery objectives are well below the range where water saving becomes a significant financial consideration for the farmer.

- The volumetric water prices required to achieve cost recovery objectives are well below the range where water saving becomes a significant financial consideration for the farmer. Some authors suggest that, to substantially reduce demand, volumetric prices would need to be 10 to 20 times higher than the price needed for recovery of the full supply cost.

- In public schemes, the objective of full cost recovery has been difficult to achieve in practice. In the overwhelming number of cases, charges do not cover even annual O&M costs. The literature refers to various institutional and political factors that hamper full cost recovery in a wide variety of countries, including:
  - The lack of political will to impose higher charges on farmers.
  - Unwillingness to reduce costs by slimming down over-staffed government agencies.
  - Lack of motivation on the part of agencies responsible for fee collection, as fees return to the treasury, rather than the agencies, with no direct link between revenue collection and operational funding.
  - A vicious cycle of low O&M expenditure leading to poor performance and an increasing reluctance on the part of farmers to pay for an inadequate service.
  - Insufficient resources - time, money, training - given to planning and implementing cost-effective charging mechanisms.

A series of case studies were undertaken in India, Pakistan, Nepal, Macedonia and Morocco. The main conclusions of those studies are as follows:

- Farmers' willingness to pay, rather than their ability to pay, is the primary determinant of whether costs are recovered.

- Willingness to pay is greatly influenced by:
  - The equity and quality of the irrigation service.
  - Farmers' perceptions of the transparency and fairness of the billing system.
  - The political will to set charges at adequate levels and enforce their collection.
  -

- The "affordability" of a given charge must be assessed on the basis of the benefit derived from the service, and overall income. Farmers with a total income of only a few hundred dollars per year will find it difficult to pay even 5 - 10% of that income in fees. Poverty concerns suggest that governments may need to continue to subsidise O&M costs in such conditions. However, the studies demonstrate that in the majority of cases, it is reasonable to use irrigation service charges as a tool to recover at least annual O&M costs. Such charges constitute a relatively small proportion of the benefits derived from irrigated cropping.

- Where recovery of O&M costs was being achieved, charges per unit water delivered varied widely. In the Gujarat study, where farmers irrigate from collectively owned, private wells, charges exceed US¢ 5/m^3. The well-managed, surface schemes of Morocco charge fees of at least US¢ 2/m^3, equivalent to as much as 17% of net income, and these are successfully collected. In Haryana full recovery of O&M requires an irrigation charge equivalent to only US¢ 0.11/m^3, due to the simple method of water control (low infrastructure and staff costs) and cross-subsidies from other water use sectors.

- Payment for either 'sunk' capital costs or the depreciation of assets is only seen in the conditions of the small, privately owned, tubewells of Gujarat, where the capital payment was made at the time of construction. Raising charges to a level where capital or depreciation charges are covered may be financially viable on many schemes but achieving these goals will usually require a major effort of political will and substantial institutional reforms.

- In all the locations where water is scarce - Haryana, Pakistan and Morocco - systems of water allocation, rather than price, are used to limit consumption.

- The unit value of water (indicated by net income divided by quantity of water used) was compared with the implicit unit price of water (indicated by the irrigation charge divided by the quantity of water used). The indicator is far from perfect, but the higher the value:cost ratio the greater will be the demand for water, and the larger the increase in price required to significantly influence demand. The indicator varied from a factor of 3:1 for the tube-well users of Gujarat, to a factor of about 180:1 in the surface systems of Haryana. A 10 to 20-fold ratio between value and cost is seen in the other cases, consistent with values found in the wider literature review. Such a difference between benefits and costs of irrigation to farmers indicates that prices would have to be increased very substantially, *and actually be imposed*, in order to have more than a marginal effect on irrigation demand.

1.6 USING THE GUIDELINES

Figure 1 is divided into five steps, corresponding to chapters of the Guidelines. Chapter 2 sets out the importance of understanding the *policy* environment, i.e. the political and legal context, within which irrigation charging is to occur. It prompts the reader to identify existing policy positions or to seek clarification where policy is not clear. Chapter 3 takes the reader through the steps of specifying and prioritising the *objectives* that irrigation service charges are to serve, as it is only when the objectives of a charging policy are clear that the mechanism or *basis* for charging can be selected. The chapter also points out the importance of making a realistic analysis of any existing charging system. Where an existing system fails to meet its stated objectives a critical diagnosis must be carried out to identify the reasons for failure and their implications for any new system of charging.

Chapter 4 analyses the relationship between specified charging objectives and the various charging mechanisms that can be applied.

Only after these fundamental steps have been adequately addressed can the more pragmatic questions of '*how much and whom to charge*?' be quantified. Chapter 5 provides guidance and examples on quantifying and updating charges and the distribution of charges amongst different types of user. Finally, chapter 6 provides guidance on the practical *implementation* of a charging system – measurement of water volumes or crop areas, how and when to deliver bills, and incentive and sanction systems that may contribute to high levels of fee recovery. The detail of implementation will be determined by local customs and resource availability but it is vital that the resource needs of any charging system are realistically assessed and provided for to prevent the system falling into decline and disrepute.

Annex A of the guidelines provides a series of checklist tables that lead the user through the logical steps set out in the body of the text. It is recommended that those checklists provide a working basis when applying these guidelines to design or revise an ISC system.

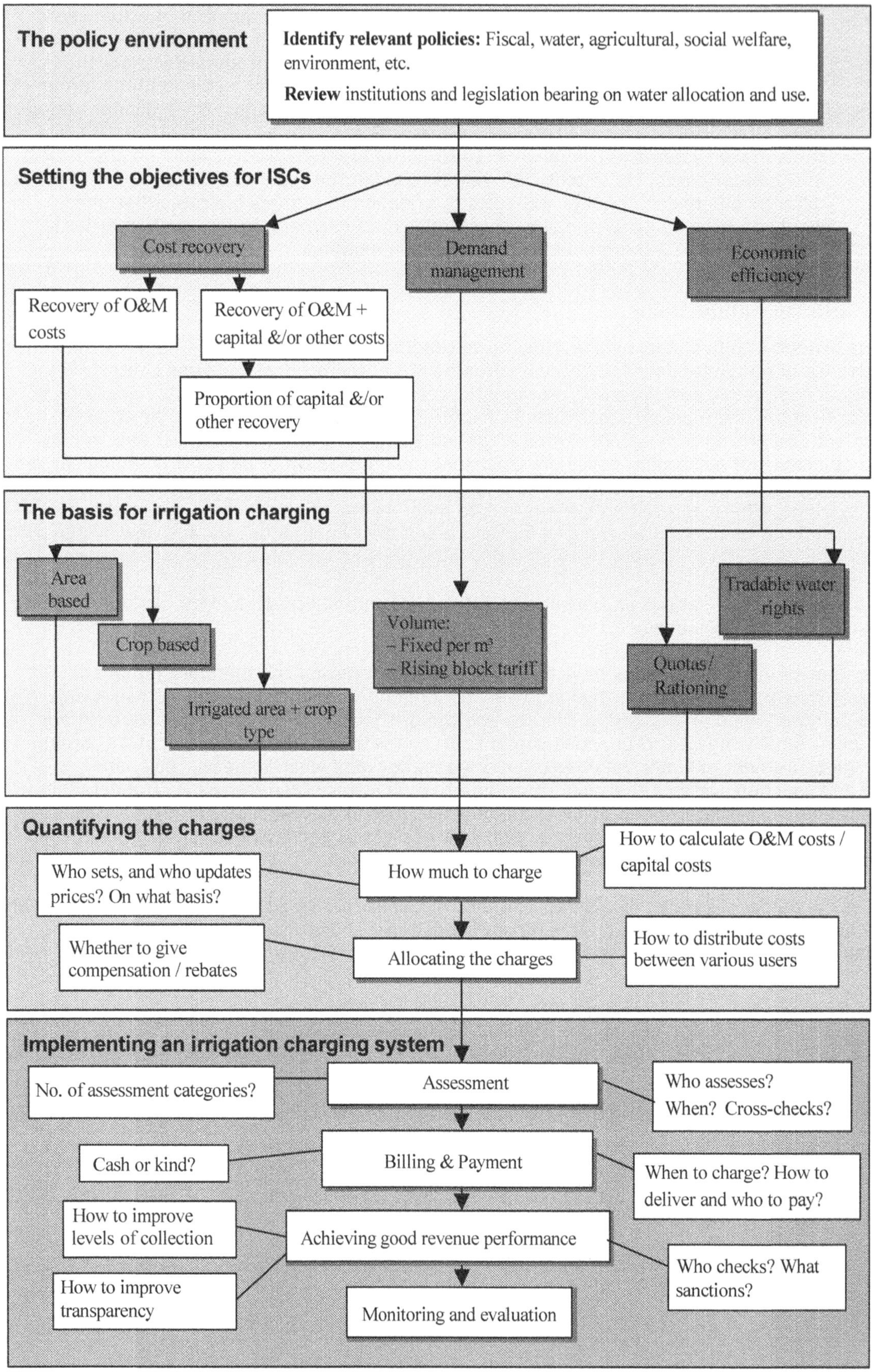

Figure 1 Steps in designing or reforming an irrigation service charge system

2. The policy environment

Existing policies in irrigation and other sectors will influence and even specify the objectives to be met through ISC, and the means that can be employed to meet those objectives. In designing or reforming ISC it is therefore essential to review all relevant policies and their implications for irrigation charging to ensure that ISC proposals are consistent with policy and draw strength from policy support. Charging for irrigation services is inevitable politically contentious and using whatever support is available from existing policy – and avoiding as many conflicts as possible – will increase the chances of success.

The elements of the policy environment can broadly be classified into strategic policy – national guidance on the role and purpose of charging – and implementation policy which sets out important practical aspects of ISC.

2.1 STRATEGIC POLICIES

Strategic policy must include at least a general quantification of national water resources and definition of who owns them and who manages them. Typically, states retain ownership of water and the right to utilise water resources as necessary in particular circumstances. Users acquire subsidiary rights to beneficial use, either through tradition or through licensing procedures. The definition of ownership and right to use leads to definitions of responsibility in development and management of the resource, and associated financial responsibilities – who pays for infrastructural investments, for maintenance of facilities, and for the associated management activities. The costs incurred in development and management are often clearly attributable to a beneficiary group – for example, the costs of drilling a private well fall directly to the owner. Other costs are less easily assigned – for example, dam safety programs directly benefit the farmers receiving irrigation from the dam concerned, as well as downstream facilities threatened in case of failure, and the population at large who benefit from agricultural production and provision of domestic water supplies. All costs associated with water resources development and utilisation must be assigned to individuals, groups, agencies, or the government.

Some countries – for example, Morocco – have specified formulas detailing the precise basis for charging for irrigation services, varying by source of water and cost of system construction. Other countries have clear but less specific policies – for example India's central government policy in the 1980-90 period was to provide storage facilities at state expense, share conveyance system costs between irrigation and non-agricultural users, and allocate distribution costs to irrigators. While useful, these policies left much to be interpreted (what proportion of capital to recover at what rate of amortisation; how precisely to "share" costs between users) and since implementation of cost recovery was a state government issue, the actual outcome was highly variable. Other countries have much vaguer policy guidance – charges should be affordable, costs should be recovered (What is affordable? Which costs to recover?). Context and timing are additional issues: for example, over the last decade, revised pricing policies for energy have rendered many lift irrigation schemes in former Soviet Republics financially unviable. Yet many thousands of people were settled in the areas served by these systems and continue to depend on them for their livelihoods: clearly, any new policy of financial self-sufficiency must in such circumstances be related to the reasonable rate of adjustment of the affected communities, while also protecting the viability of energy supply agencies.

Sectors that may have strategic policies that will influence irrigation water charging are listed below in Table 2.1.

Table 2.1 Policies affecting the design of ISC

Sector	Policy content
Fiscal/legal	• Sources of funding for water services development. • Proposed levels of cost recovery on public infrastructure and services for replacement as well as O&M costs in various water supply sectors • Policy on cost sharing where different sectors share common infrastructure • Ownership of facilities, public and private responsibilities in maintenance.
Water	• Inter-sectoral priorities for water allocation. • Procedure for definition of water rights at state, sector, project and intra-project levels - including non-consumptive uses such as hydropower, fisheries and navigation. • Policies on groundwater management and exploitation. • Roles and responsibilities of water service entities. • Legal status, roles and responsibilities of irrigation water users.
Agricultural	• Policies promoting or discouraging production of particular crops. • Food security policies and agricultural trade issues.
Social Welfare	• Policies favouring defined social groups regarding access to water for agriculture, including priorities between urban and rural communities.
Environment	• Pollution control or pollution charging systems.

Some policy areas, for example, in social welfare, or policy on water allocation within agriculture, are in a sense optional: if they exist, they will influence the design of ISCs; if they are not in place ISCs can be formulated in their absence. However, policies in the fiscal/legal and water areas are essential to allow specification of ISCs and the associated service that irrigation beneficiaries are to be provided with. To the extent that there are gaps in these areas the ISC system will need to specify, or presume, a policy, which in turn will require political and legal endorsement.

The first, and often most controversial, policy requirement is the specification of which costs are to be recovered and whether full or partial recovery is to be achieved. Key issues are the extent to which historic investment costs are to be repaid - although "full cost recovery" is widely advocated in theory, most publicly developed systems have not pursued this objective even in the transition to financially independent operation (see Box 2.1).

Although India's record in cost recovery for irrigation and other water services is not strong, the policy guiding the recovery of costs was clearly specified some decades ago. It requires that the cost of storage facilities be borne by the State, the cost of conveyance facilities be allocated among the various beneficiaries (agriculture, domestic and industrial - recognising abilities to pay), and the cost of distribution facilities be borne by the direct beneficiaries - usually irrigated farms. Such guidance, though lacking in some details, provides a clear basis for the formulation of consistent rules for allocating and recovering capital costs.

Most countries explicitly give priority allocation of water to human needs, followed by industrial, irrigation and environmental requirements. Depending on the priority given to irrigation (which may be further divided - for example, Florida gives priority to citrus trees, followed by vegetables, and finally field crops) the appropriate charge for services to each of these uses would reflect the relative certainty of supply.

Box 2.1 Defining the value of existing capital assets

The privatisation of the UK water industry transferred a previously state owned and operated set of infrastructure to private commercial management. Asset valuation was a key issue, since the operating companies' revenues were set based on securing a reasonable return on assets. Since most of the assets were more than a century old and buried deep underground, valuation was based not on estimates of cost of construction or replacement, but rather on the likely income the assets could generate. This in turn was based on historic (subsidised) revenues paid to the state operators, resulting in a substantially lower valuation of the assets.
Similarly, the recovery of costs in the Murray-Darling basin is largely unrelated to the historic cost of assets. In both cases, however, current and future investment costs must be fully recovered, and as investment continues, the original value of assets becomes a declining component of the nominal asset base.

Another difficult area is the extent to which non-financial costs (social, environmental) should be incorporated into ISC. Again, while theoretical studies have identified and attempted to quantify such costs, in the practical world, objectives such as sustaining wetlands or ensuring that female headed households have equitable water allocations are achieved by physically allocating water to such purposes rather than through indirect charges, taxes or subsidies designed to result in a particular allocation pattern.

The remaining components of ISC requiring policy definition are clearer, relating to actual current costs of operation and maintenance, the treatment of unpredictable events - failure of a structure, storm damage - and the costs of replacing and upgrading facilities. If financial independence is to be achieved, such expenditures must be met, and the charges in any given period will often significantly exceed expenses in that period. How such funds are kept, by which organisation (if there is more than one beneficiary of the system, and more than one level of organisation in the institutional arrangement) all require detailed specification.

If there are major gaps in policy implementation - for example, if water distribution is currently determined in ways that are unrelated to stated policies, or current ISC are not collected - decisions will be required regarding priorities for intervention. It may be preferable to address the policy failures before proceeding to design ISC.

If there are major gaps in policy implementation it may be preferable to address the policy failures before proceeding to design an ISC.

2.2 IMPLEMENTATION POLICY

Implementation policy will be highly dependent on the institutional framework for managing irrigation. Where, as for example in many South Asian countries, the agency managing irrigation is a department of government, the whole process may be concentrated within government. Even then, more than one government department may have responsibility - thus in northern India and Pakistan, the amount of charge to be paid is computed by the Irrigation Department, while billing and collection is done by the Revenue Department. Arrangements are made more complex when farmer organisations are created in the process of management transfer. There can easily be three institutional levels - farmer group, operating the lowest level of management; federations of farmer groups operating to manage and share water among the farmer groups, and irrigation agency - usually still a government department. Where such arrangements are in place, policy decisions are required as to the financial responsibilities at each level, the revenue sources available to each level, and the sanctions and enforcement procedures between levels - based on defined service agreements at each level. Box 2.2 illustrates the need for policy and legislation to strengthen institutions in Pakistan.

Box 2.2 Irrigation Service Charge (ISC) in farmer managed systems, Pakistan

Over the last two decades, Pakistan has struggled to improve the recovery of abiana (irrigation service charge). Abiana is supposed to cover the service costs of the distributary and upstream system (dams, head works and main canals) but at the end of 1999, the cumulative arrears in payment was about 44% .

In Sindh, Farmers Organizations, introduced under institutional reforms, are required to collect abiana to cover services including drainage. The organizations should retain 40% of total revenue and pay 60% to the service provider - the Area Water Board.

In the first year of operation (2001/02), virtually all dues were paid to the Area Water Board. However, there were substantial shortfalls in the proportion of the collections intended for the organizations themeselves. In effect, many farmers paid only 60% of their dues, with the result that the organizations struggled to carry out operation and maintenance of the minors under their control.

To improve the situation it is being proposed that, amongst other changes Farmers Organizations should have magisterial powers to collect abiana, an authority currently vested solely in the Revenue Department.

2.2.1 Who assesses and collects the Irrigation Service Charges?

Whatever the basis for the ISC - volume of water delivered, crops grown, area irrigated - a formal record of the service must be kept in such a way that the service provider and the service recipient agree on what payment is due. Where service delivery is multi-tiered (agency, Water User Federation (WUF), Water User Association (WUA), etc) the basis for payment and the collection agency may be different at each level - but at every level responsibilities must be clear and the action to be taken in case of default must be agreed, recorded and enforceable.

2.2.2 On whose behalf are Irrigation Service Charges collected?

Irrigation charges are usually collected either on behalf of a general tax authority or the irrigation service provider. In the first case, irrigation charges are seen as one tax amongst other taxes and the receipts added to a general public revenue department. It is not uncommon that irrigation charges are then collected with other taxes. Under this arrangement, the irrigation provider is generally funded like any other public agency and the amount of funding provided by government to the agency may be more, less, or the same as the amount collected through ISC.

> ***When irrigation charges directly finance the irrigation service provider, the relationship between the provider and water users is clear and the charges become a 'service fee'. The water users have a direct stake in the services of the irrigation provider and the irrigation provider depends on their financial contributions. This gives a wide range of opportunities to define and control service levels and improve accountability.***

Where irrigation charges are collected on behalf of the irrigation agency, the agency usually has a greater degree of financial autonomy because it is at least partially self-financing. It may still engage a third party to do the collection - which may even be a tax department - but the net proceeds contribute directly to the agency's own budget.

> **Box 2.3 Beneficiaries become customers and cut their costs - an example from Mexico**
>
> When irrigation systems in Mexico were turned over to farmer management (with full recovery of O&M and improvement costs from farmers) the former government agency was essentially disbanded at field level. The WUAs decided whom to hire from the former agency staff. Less than half the former payroll was re-hired, but at considerably higher wage rates. Farmers' experience of the honesty and dedication of the staff were major determinants in this process.
>
> Source: Sam Johnson III - personal communication

When irrigation charges directly finance the irrigation service provider, the relationship between the provider and water users is clear and the charges become a 'service fee'. The water users now have a direct stake in the services of the irrigation provider and the irrigation provider depends on their financial contributions. This gives a wide range of opportunities to define and control service levels and improve accountability. At the same time, it makes the overall organization of ISC collection more complex and wide-ranging - especially if the arrangement is multi-tiered, involving WUAs and or WUFs. This is particularly important when irrigation providers make the transition from a public agency to an autonomous body. As this happens, the relationship between the beneficiaries and the service agency or agencies changes profoundly. Beneficiaries become customers paying the proper price for the service; further, the customers may have the right to see how the payments they make are utilized, to influence the selection of agency management (see Box 2.3) and to make critical choices regarding which management and investment tasks are undertaken, by whom, and when. In response to these changes, service providers are likely to address new issues. In particular, maintaining accounts so that the costs of service provision are clearly identifiable and also more generally in "customer relations" so that the beneficiaries who now fund the service are made aware of what decisions are taken and why.

There are clear benefits to the direct relationship between a self-sufficient agency providing irrigation services, and beneficiaries paying an appropriate charge (and aware of what the charge is being used for). The agency will be accountable to those who pay (rather than to central government funds), and the beneficiaries will feel empowered to monitor performance, and will benefit if the service is properly funded. However, while almost all privately owned and managed systems operate on this basis, many large public systems operate entirely as government-provided services with no link between payment for service and costs of providing the service.

There are intermediate situations (Box 2.4), and often a switch to the independent, fully accountable model should be undertaken with care: if the institution collecting and spending ISCs is not sound, the potential for misuse of funds and discrediting of the "payment for service" model is high.

Box 2.4 Financially Independent Agencies - Intermediate models

Most surface irrigation in India is essentially a government service. In Haryana, steps have been taken to ensure that the overall recovery of costs is adequate to cover the associated costs. The sensitivity of politicians to the level of charges resulted in significant pressure to reduce the costs of the service through savings in staff. Although the objective of cost recovery has been met, the collection of charges remains with the Revenue Department, and funds go directly to the central revenue account - not to the Irrigation Department.
In Morocco, the system is somewhat more decentralized - though not fully.
Revenues are retained by the local ORMVA (irrigation service agency), but shortfalls are covered by subventions from the central government.

Source: Case studies

2.2.3 Who to charge for irrigation services?

The usual choice is between the landowner or the water user, who may not be the same. If the purpose of charging is to generate revenues, the charging agency (whether government or WUA) will primarily consider administrative efficiency and the scope for sanctions in the case of default. If the objective is to send signals on the value of water, charging absentee landowners may be ineffective. On the other hand, the administration needed to charge tenant water users may be more complex and the transaction costs higher because an additional registration of the user (separate from the landowner) is required, and enforcement of payment cannot be linked to charges on the land.

Another important consideration is whether, and how, to charge non-agricultural water users. Many irrigation systems are multi-functional and serve a wide diversity of interests within irrigation systems - domestic water supply, fisheries and other enterprises. Some of the non-agricultural uses are of the nature of a public good and would justify the need for public support. However, most non-agricultural users derive a much higher value from water services - provided the service gets priority in times of shortage and is continuous throughout the year. Box 2.5 provides an example of this.

Box 2.5 Differential pricing for non-irrigation users

Under a World Bank-funded project in Haryana, India, the state government agreed to charge water users adequately to recover the full costs of operation and maintenance. Water resources in the state were controlled by the Irrigation Department, who had paid little attention to the potential for cost recovery from non-irrigation users (all towns and villages in areas with saline groundwater - about a third of the State - utilized treated canal water for domestic and other non-agricultural uses). Reflecting the higher level of continuity, security and reliability of supply afforded to non-agricultural users, the State now recovers some two thirds of the overall cost of water service provision from non-agricultural users while their total water allocation is less than 10% of the total distributed. The average cost of water delivery in the state is US¢ $0.13/m^3$, while agricultural users are charged only US¢ $0.05/m^3$.

2.2.4 *Social objectives and service charges*

Subsidized irrigation services are often proposed on social grounds. For instance, the original development of irrigation systems in much of South Asia was justified as protection against drought; protecting the local population from famine and providing increased food security to the nation. Such considerations serve to insulate the beneficiaries from arguments that they should pay the full cost for the service they receive, and there are situations - certainly in periods of significant economic transition - when such considerations are powerful. However, where these arguments are accepted, the issue that must still be addressed is financial sustainability: if beneficiaries do not pay for the service they receive, either the government provides the shortfall or the facilities will deteriorate. The key to poverty alleviation is to enable farmers to grow crops that allow them a decent living. The performance of the irrigation system should be evaluated - and funded - on this basis. Box 2.6 provides examples of subsidies offered to poorer, small farmers in Bulgaria and Slovakia.

Box 2.6 Exempting small irrigation users - cases from Eastern Europe

- Exemption of payment in notified areas for usage of less than 10 cubic meter/day or less than 0.2 l/s (Bulgaria).
- Exemption of payment for farms of less than 0.2 ha for the first 3000 cubic meter per ha per month (Bulgaria).
- Payment exemption for usage of less than 1,250 cubic meters per month (Slovakia).

2.3 CHAPTER SUMMARY

See Table 2.2.

Table 2.2 The policy environment – chapter summary

Step	Actions/Issues	Recommendations / notes
Strategic policies	• Quantify national (State/regional) water resource base and its ownership. • Quantify costs of resource development and use and how they will be recovered. • Specify ownership of infrastructure and responsibilities for costs of maintenance, upgrading and replacement.	Determine who pays, what amount and over what time period.
	Ensure consistency of charging policy with policies in other sectors.	Integrate irrigation policies with those of other water users and within wider social policies.
Implementation polices	Define: • How and at what level, service is defined and measured. • Which agency will collect charges	Formation of water user associations avoids need for bulk service provider to deliver quantified level of service to many thousands of individual users. WUAs monitor service at point of transfer and are accountable for payment to bulk supplier.
	Define whether charges return to central treasury or are retained on the scheme.	Charges should go directly to the irrigation service provider. The relationship between the provider and water users is then clear and the charges are a 'service fee'. The users have a stake in the services of the irrigation provider and the provider depends on their financial contributions. This gives a wide range of opportunities to define and control service levels and improve accountability. If charges go to the treasury a mechanism is needed to relate budgetary allocations to services and levels of fee recovery.
	Define if charge is levied on user or landowner (tenant or landlord).	Charging an absentee landlord may ensure cost recovery but will not influence water use behaviour in the field – therefore decision must be based on the objective of charging.
	Consider the need for subsidised irrigation services to support the livelihoods of rural poor.	If beneficiaries do not pay for the service they receive, the government must fund the shortfall with subsidies that are in line with declared social policy. Failure to provide adequate budgetary support leads to deterioration of infrastructure and reduced benefit to intended 'beneficiaries'.

3. Setting the objectives for Irrigation Service Charges

The review of national or regional polices in water-related fields will identify key policies which will shape the overall purpose and some of the detail of ISC. In addition, problems within the irrigation sector may be addressed in the design or reform of ISC and water allocation procedures. Figure 2 provides a summary of possible problems and their potential causes that may be linked to underlying policy weaknesses or the implementation of irrigation charging. Where the concern is to reform an existing ISC system the figure provides a simple checklist of the issues to consider in diagnosing that system.

The three most widely stated objectives of irrigation charging are:

1. To achieve a specified and consistent level of *cost recovery* from users.
2. To provide an incentive to irrigators to *reduce water consumption* (demand management).
3. To *increase the productivity* of water at the individual user level, or through transfers to more productive users or uses – the goal is to increase the level of economic benefit derived from a given amount of water.

Where the design and implementation of a charging system delivers one or other of these objectives it will address at least one of the problems identified in Figure 2. Unfortunately, policymakers often express an interest in achieving all of these objectives through a single charging mechanism, but as chapter 4 will show, the ISC features that serve one purpose often do not serve others – or indeed may actually conflict. Indeed, it is important to note that an irrigation service charge – including volumetric charging – may not be the most appropriate or effective tool to realise the second or third objectives, as chapter 4 will demonstrate. The most important decision at this stage is thus to prioritise objectives. As the options for achieving the most important objective are reviewed, their potential relevance – or otherwise – to secondary objectives can be assessed. If more than one objective must be addressed, it may be necessary to introduce additional components to the ISC. Most commonly, where the quantity of water available varies sharply from season to season, while costs of service provision vary much less, a two-part tariff is required to meet objectives of relating the charge to the quantity of water delivered, and ensuring stable revenues.

The issue of seeking multiple objectives is complex: "demand management" and "stable revenues" are two common and appropriate objectives with significant potential contradictions. The price that will result in a particular desired level of demand will vary from year to year, and there is no reason at all to expect that that price, multiplied by the available water, will correspond either to the revenues required to meet cost recovery objectives, or that it will be stable.

It is important to recognise that an intervention may *contribute* to an objective (thus an increase in charges may result in some degree of reduced demand) but not fully meet a precise objective such as reducing demand to a specific level. The benefit of reduced demand should not be discounted even if minor – but neither should it be overstated to suggest that the problem of excess demand is automatically solved by a price intervention.

It is also important to stress that to achieve any of the three objectives it is essential not only to select the appropriate mechanism, guided by Chapter 4, but also to give adequate attention to effective field implementation, using chapters 5 and 6 as guides.

Figure 2 Typical problems arising in irrigation charging systems

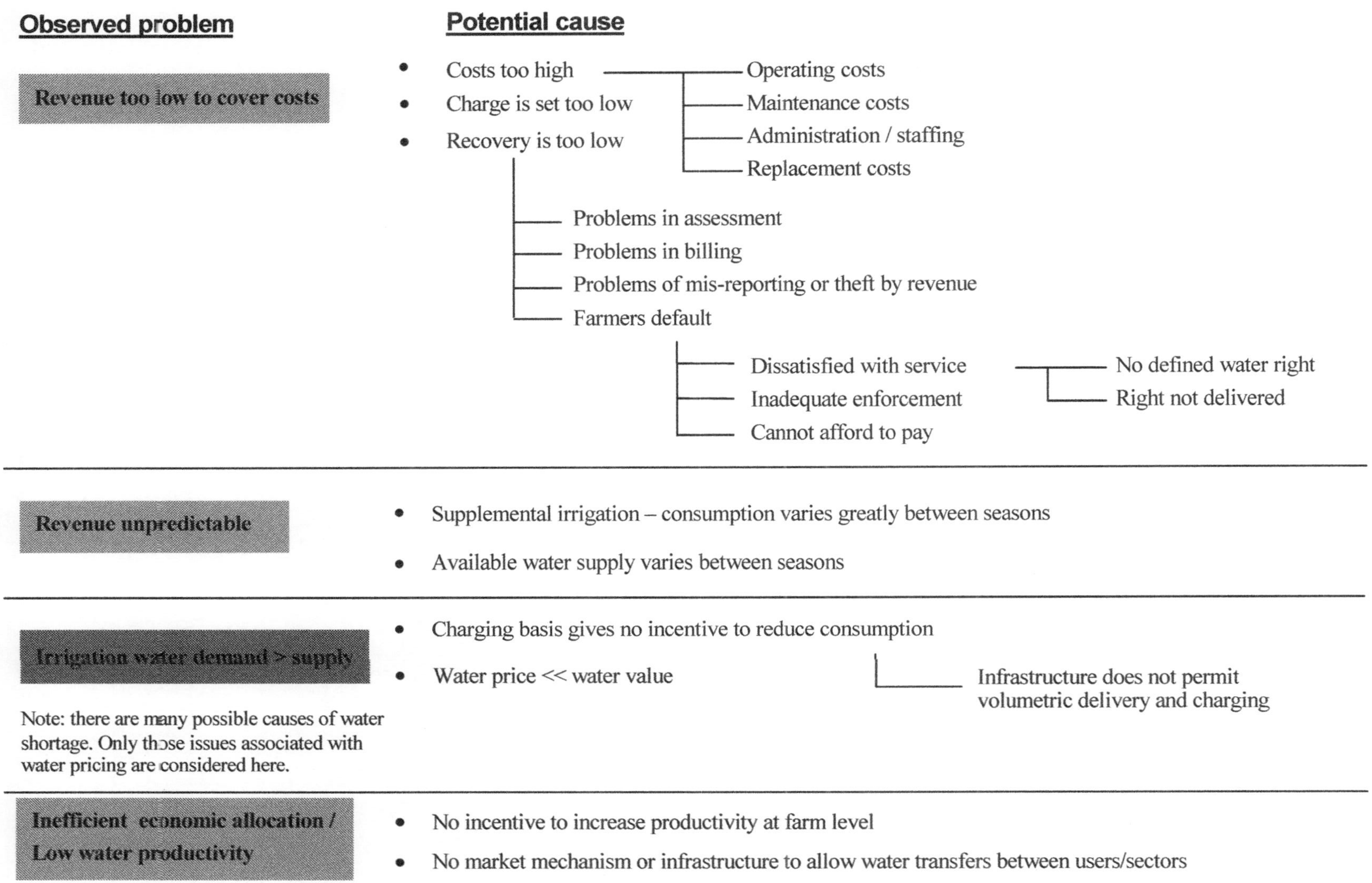

3.1 COST RECOVERY

Adopted policies must define the extent to which cost recovery is required. The ongoing financial viability of an irrigation system requires that income from all sources must equal day-to-day operating costs, replacement costs for major facilities, and other investments. As pointed out in the previous chapter (Section 2.2), financial viability does not demand the recovery of all previous capital expenditures or accumulated debts from the irrigators. In some irrigation systems, governments have decided to provide services to otherwise backward areas to stabilise existing or settler populations. In such circumstances, irrigation charging policies are shaped by wider political and social policy objectives, and the financial 'viability' of the schemes must be ensured through subsidies. Difficulties occur when charges are inadequate to meet financial sustainability needs, *and a compensating subsidy from government is not provided.* In such cases, in order to provide the political basis for raising charges to ensure financial viability, policy decisions allowing increased charges, or providing the necessary subsidy, are a precondition to formulating ISC to meet the specified cost recovery objectives.

Box 3.1 demonstrates the danger of a sudden policy decision to recover all past costs from the irrigators.

Box 3.1 Macedonia – recovery of past debts reduces revenue

In Macedonia, policy requires full cost recovery including the original investment costs, interest, and accumulated debt. The resulting charges are so high (and the agricultural economy so unstable) that many farmers opt not to irrigate. As a result, the overall fee collected is reduced and the financial situation deteriorates further – an entirely unsustainable situation.

Where responsibility for the O&M of assets is transferred to the users but ownership of the assets remains with the state, users may be reluctant to pay for future replacement costs. Further, when users are responsible for "routine" maintenance, while government undertakes to fund major works, an incentive is created to allow minor problems to escalate. While this issue is not directly related to ISCs, it illustrates the need to look out for unexpected "incentives" in any payment scheme.

Recovery of annual O&M costs, as a minimum, is affordable on almost all schemes, provided the basic preconditions concerning the implementation of charges, set out in Section 4.4, are met. However, it is important that the following characteristics of annual O&M costs are recognised when designing an ISC:

a) Costs will vary widely between schemes according to their water source, age and topography, amongst other factors. Thus, the application of a single level of charge at a state or regional level will seldom reflect the actual O&M costs of many schemes in that area. If the actual O&M costs are to be recovered, individual schemes must be permitted to fix their own charges.
b) The greater part of annual O&M costs are completely independent of the area irrigated within a season, the number of irrigations made or the volumes of water released – the costs of staff and the maintenance of infrastructure are independent of these variables. A charging system based solely on these parameters may generate a highly variable income stream with very low income in years of drought or where rainfall makes irrigation unnecessary. Use of a 2-part tariff, as described in Chapter 4 (4.1.4) will greatly reduce such annual variation in revenue and permit more rational annual budgetary planning.

Where the primary objective is to bring revenue and expenditure into balance it is essential that irrigators see that efforts have been made to reduce costs to an 'acceptable' level. The tendency of public agencies to retain large numbers of poorly motivated and poorly resourced staff as a means to provide employment is widely recognised. Irrigators will be more likely to accept increases in irrigation charges that bring revenues and expenditures into line when they see tangible evidence of efforts to reduce over-inflated costs.

3.2 DEMAND MANAGEMENT

Demand management has driven much of the recent discussion of ISCs, yet the concept is often not well specified. "Demand management" in the context of irrigation was formulated as a phrase to distinguish between supply-side interventions (construction of new dams, for example) and demand-side interventions aimed at "managing" within the existing, sustainable, supply. In some contexts, it would be more accurate to describe such interventions as limiting the use or consumption of water rather than managing demand. The demand for scarce water remains well above the volume used if quotas are enforced, or irrigated areas are limited. It is usage that is being controlled rather than demand - but in this text the now current terminology "demand management" includes:

- Any intervention or incentive to avoid absolute wastage of water such as letting irrigation water run to drains.
- Setting a charge that makes some uses unprofitable, hence reducing demand.
- Setting a charge designed to achieve a specific reduction in demand.
- Allowing the price of water to rise to a level that brings supply and demand into equilibrium - the market-clearing price.

In fact there is overlap between this range of objectives and the objective of increasing productivity because any intervention that releases water from a low productivity use can allow the water thus released to be more productively used. Which objective is sought has fundamental implications for ISC design.

In Annex B it is shown why an incentive to reduce consumption is provided only where the charge is directly based on the volume of water consumed or a proxy for it, such as the duration of irrigation supply or number of irrigations. Water charges linked to crop types and their relative water demand will only reduce total consumption when the price is set at a level that makes certain crops less profitable than others consuming less water. At this point farmers may change their crop, although factors other than the cost of water, such as prices and marketing and the resulting crop profitability, will always influence cropping decisions.

Since the value of water to irrigators often exceeds the existing price, or the price required to recover all annual O&M costs, by a factor of 10 to 20 times or more, a significant reduction in consumption may only be achieved with a price rise that would be considered politically unacceptable and socially undesirable. However, volumetric charging will generally discourage profligate use or wastage of water and promote higher water productivity - more production from the same volume of water applied to the field.

Whether the actual volume of water consumed by evapotranspiration falls is a different question; advanced irrigation technologies generally improve the delivery of water to crops and in consequence increase evapotranspiration by the crop, increase the crop yield, and often increase the productivity of water. However, this can paradoxically lead to an increase in water demand - especially if water is scarce and land is plentiful - because once the "better" technology is in place, the farmer will spread the water delivered over a larger area, increasing consumptive use and reducing return flows. Thus, it is important to be very clear about the outcome that is sought before using charging as a tool to modify farmers' behaviour and demand for water. If farmers have the opportunity and incentive (prompted by the marginal cost of water) to increase the productivity of both land and water, this may result in a rise, rather than fall, in the volume of water consumed through evapotranspiration. This may reduce return flows and actually increase negative impacts on downstream users. Furthermore, if the productivity of water is increased, the incentive to increase consumption (by diverting more from rivers, or pumping more from groundwater) is similarly increased as long as the farmer has available land.

3.3 INCREASING THE PRODUCTIVITY OF WATER

Water, like any other scarce resource, should be used for the purpose giving the highest added value to whatever objectives society, through politics and laws, has set. Usually the dominant objective is economic productivity. We have noted above that interventions to manage demand may well have positive productivity impacts - but it is useful to keep in mind the separate objective of increasing productivity.

In irrigation, increased water productivity can be achieved at the level of the individual farmer by encouraging waste avoidance and promoting crops that give good returns to water. The objective is more broadly achieved if water is reallocated from less productive to more productive farmers, or, at the extreme, reallocated from less to more productive economic sectors.

This implies reallocation of water among users and uses. It requires an incentive structure to drive such reallocation (assuming administrative fiat is not the chosen means). Pursuing this objective through purely "market" means incurs all the practical difficulties listed above in respect of volumetric water charging. Furthermore, transfers of water usually have third party impacts (on groundwater recharge, in-stream flows, distribution of losses) that require well-organised support and administration systems. The most practical way of achieving the goals of intra and inter sectoral transfer appears to be through tradable water rights. This allows trading of water in bulk, and only involves "willing partners" in the trading and pricing process, rather than enforcing market pricing on all users.

3.4 CHAPTER SUMMARY

See Table 3.1

Table 3.1 Setting the objectives for Irrigation Service Charges – chapter summary

Step	Actions/Issues	Recommendations / notes
Prioritise objectives	• Cost recovery	Income from all sources must equal day-to-day operating and maintenance costs plus replacement costs for major facilities. If it is not feasible to recover these costs fully from users, government must provide a clear subsidy for the shortfall. Sustainability does **not** require recovery of non-replacement capital costs as, by definition, it is not anticipated that these capital items will be replaced.
	• Demand management a) Providing an incentive to reduce waste. b) Bring supply and demand into balance.	There are **major technical barriers** preventing provision of variable volumetric supply to individual users in many irrigation systems in the developing world. **It is strongly recommended that price is not used** to try to bring supply and demand into balance. This is more simply achieved through allocation mechanisms.
	• Increasing the productivity of water – reallocation to more productive users	Volumetric pricing may stimulate some increase in productivity but seldom leads to water transfers out of agriculture to other sectors. Tradable water rights can facilitate such transfer (See 4.1.6).
Review existing charging system in the light of defined objectives	• Identify weaknesses and define needs for change	See Figure 2.

4. The basis for irrigation charging

In the previous chapter, the objectives of ISCs were reviewed, focusing on the three most common - cost recovery, demand management, and increasing the productivity of water. The first of these objectives can be met through any form of charging, provided the rates are set at a suitable level. In this chapter we review the various bases for charging for water or controlling how water is allocated, and assess their potential contribution to meeting demand management and productivity objectives. They are considered both from the point of view of the user, i.e. how they will influence the behaviour of water users, and from the perspective of an irrigation agency. In this second case we consider the agency's preference for a stable and predictable revenue stream and a tool that is easy to administer.

4.1 CHARGING SYSTEMS AND WATER ALLOCATION PROCEDURES - IMPACTS ON WATER USERS AND IRRIGATION SERVICE AGENCIES

Table 4.1 sets out the range of irrigation service charges and water allocation procedures (Types), summarising the potential of each to contribute to the objectives of encouraging productive use and reducing water demand by the water user. Types 4 and 5 are not charging mechanisms and used in isolation they do not generate an income stream. However, they may be considered as 'economic instruments' as they provide a means to control the allocation and use of a scarce resource. More importantly, if they provide a more effective and transparent means of achieving demand management and the allocation of water to the most productive users and sectors than direct charging systems, then they should be considered by practitioners and hence they are included these Guidelines.

Table 4.2 presents the same five types of irrigation charge or allocation method from the perspective of the irrigation service provider, summarising the effect of each 'type' on the predictability and stability of revenue and its ease of administration. These are normally the priority concerns of a service provider, often taking precedence over the need to improve productivity and limit demand.

Tables 4.1 and 4.2 show a degree of conflict between an agency's likely priorities and the types of system that can be used to provide productivity incentives to farmers and encourage water saving. The systems that rank highest from the agency's perspective - predictable income and simple to administer - are least effective in encouraging productivity and reducing demand for water.

If it is necessary to move beyond cost recovery to achieve the additional objectives of demand reduction and increased water productivity, then it may be better to combine an area based charge to recover costs with a quota that limits demand and prompts farmers to increase the productivity of water.

Where it is necessary to move beyond cost recovery to achieve the additional objectives of demand reduction and increased water productivity, then in many settings it may be better to combine an area based charge to recover costs with a quota that limits demand and prompts farmers to increase the productivity of water. This recommendation runs contrary to many of the recent calls for the use of volumetric water charging as a tool that can deliver all three objectives. The following sections set out the strengths and weaknesses of the five different types of charging and water allocation tool.

Table 4.1 Irrigation charging and water allocation procedures – their influence on water productivity and demand.

Type	Detail	Impact on Productivity	Impact on Demand	Can assure supply/ demand balance?
1. Area-based	1.a) A fixed rate per hectare of farm, unrelated to the area irrigated, crop grown or volume of water received. This type of charge is commonly part of a "two part" tariff – designed to cover the fixed costs of the service.	None	None	No
	1.b) A fixed charge per hectare irrigated, and not related to farm size, type of crop grown, or actual volume of water received.	None	Small	No
2. Crop-based	2. A variable rate per irrigated hectare of crop i.e. different charges for different crops, where the service charge is not related to the actual volume of water received, although the type of crop and area irrigated serve as proxies for the volume of water received.	Small	Small	No
3. Volumetric	3.a) A fixed rate per unit water received, where the service charge is directly related to, and proportional to, the volume of water received.	Positive	Positive	Very difficult
	3.b) A variable rate per unit of water received, where the service charge is directly related to the quantity of water received, but not proportionately (for example, a certain amount of water per hectare may be provided at a low unit cost, and additional water at a higher unit cost.) This method is also referred to as a rising block tariff.	Positive	Positive	Difficult
4. Quota or rationing	Entitlement to water is defined (absolutely, or qualified by actual availability)	Positive	Controlling	Yes
5. Tradable water rights	Entitlement to water is defined (absolutely, or qualified by actual availability) and may be sold to other users seasonally or in perpetuity.	High	Controlling	Yes

Notes on ratings: "Small" – essentially no impact, except at extreme (and unlikely) charging levels.
"Positive" – impact will be in desired direction, with magnitude dependent on level of charge.
"High" – impact substantial, independent of chosen charging system.
"Controlling" – Specifies the maximum demand that will be satisfied under different supply conditions.

Table 4.2 Charging and allocation procedures, stability of revenues, and ease of administration

Type	Description (for Detail see Table 4.1)	Stability and Predictability of Revenues	Ease of Administration
1. Area-based	1.a) A fixed rate per hectare of farm.	Good	Good
	1.b) A fixed charge per hectare irrigated.	Moderate	Good
2. Crop-based	2. A variable rate per irrigated hectare of crop.	Moderate	Moderate
3. Volumetric	3.a) A fixed rate per unit water received.	Poor	Poor
	3.b) A variable rate per unit of water received.	Poor	Poor
4. Quota or rationing	Entitlement to water is defined.	Not relevant	Variable[1]
5. Tradable water rights	Entitlement to water is defined and tradable.	Not relevant	Low[2]

Notes: 1. Ease of administration for quotas depends on the nature of quota: proportional division of water, uniformly across the irrigated area provides an exceptionally simple system to administer (Haryana); seasonally variable allocations with varying schedules of delivery (Morocco) are complex to administer.
2. Tradable water rights are complex to administer, combining the difficulties of a sophisticated quota system with the additional need to adjudicate on the third-party impacts of transfers.

4.1.1 Area-based charging

In the case of area-based charges it is important to realise that the *marginal price* of water is zero: in the absence of other constraints, the farmer takes all the water required to satisfy his estimate of the needs of the crops. Such a charging system therefore has no effect on water productivity and demand and some argue that it encourages profligate use by 'sending the wrong signals' to users – having paid for the water, the user feels entitled to take what he wants. However, an area-based charge, particularly one based on the total irrigable land holding, and not the area actually irrigated in a given season, is easy to administer and provides a stable revenue stream.

4.1.2 Crop-based charging

Under a crop-based charging system a farmer's choice of crops may be influenced by differential charges, but having chosen a particular crop mix the farmer has no incentive to save water by improved management or investment, and the charge is unaffected by the amount taken.

Crop-based charges generate a relatively predictable revenue stream to recover costs. Since the farmer is left free to select his cropping pattern, there is no reason to expect supply and demand to be equalised. The charging system will not reallocate water to more productive users or sectors, but it may provide incentives to reduce water consumption at the farm level, if the charge is sufficiently high to prompt the grower to switch from high to low water-consuming crops.

The administration of a crop-based charge requires seasonal assessment of the cropping pattern adopted by every irrigator. This is traditionally a labour intensive process which is open to abuse, but in some areas use of satellite imagery has greatly improved the ability of irrigation agencies to make this assessment (See Section 6.1.2).

4.1.3 Volumetric charging

Volumetric charges can be set to recover costs but because the volume consumed may vary significantly between seasons the revenue stream can be highly unpredictable. To minimise this an additional, fixed charge per hectare is usually needed to provide a certain minimum stable income to the irrigation service agency, (See the following Section 4.1.4 on two-part tariffs).

Under this type of charge the marginal price the farmer pays is equal to the price per unit of water - he will pay more if he takes an additional unit, less if he takes less. This pricing method is uncommon in irrigation, although it is the most common type of price in everyday life: apples, shirts and motorcars are generally all sold at a constant unit price - as you buy more you pay more. This form of pricing provides an incentive to save water that is not provided by types 1 and 2. In theory, as less productive users reduce their consumption, so as to reduce their costs, water becomes available for more productive users, thereby increasing overall productivity. It is for this reason that *marginal pricing* of irrigation water, that is, pricing based on the volume of water taken, is seen as a means of reducing demand.

However, the smooth relationship implied in Figure A1, in Annex B, where price and quantity are directly linked, requires that *each* purchaser can independently decide how much water to buy at the offered price, and that this desired level of supply can be individually delivered to the purchaser. Thus, for volumetric charging to serve as an incentive to reduce demand, the irrigation infrastructure and methods of water ordering and distribution must allow different volumes of water to be delivered and measured to individual farms. The infrastructure of many irrigation schemes in the developing world is designed to deliver water either on the basis of continuous flow or by rotation. Such infrastructure and its management are generally entirely incompatible with controlled, individualised volumetric delivery to farmers. This vital factor can make volumetric charging an impractical basis for charging on schemes in many developing countries.

> ***For volumetric charging to serve as an incentive to reduce demand, the irrigation infrastructure and methods of water ordering and distribution must allow different volumes of water to be delivered and measured to individual farms. However, the infrastructure and management of many schemes are incompatible with controlled and individualised delivery to farmers. This vital factor can make volumetric charging an impractical basis for charging on schemes in many developing countries.***

It is sometimes suggested that water be supplied on a volumetric basis to an aggregated, intermediate point such as a WUA, where monitoring of the volume delivered is relatively easy. The WUA would then take responsibility for allocation among individuals. While this approach has the advantage that within the WUA peer pressures may assist in controlling excessive demand, the question remains as to how infrastructure that has proven inadequate to make individualised and measured supplies under agency management should be amenable to such management by farmers

There is no necessary relationship between the marginal price and total charge. In particular, where rising block tariffs are applied (system 3b), it may be that low rates for substantial initial entitlements, combined with very high rates for additional water, result in a low total charge and a high marginal price. In such a case the charging agency has the opportunity to provide strong *marginal* incentives to economise on water use without an excessive total charge for the irrigation service. By contrast, a high crop-based rate (System 2) represents a high total charge with zero marginal price. As a consequence, the impact on farm incomes is high while the incentive to economise is minimal.

4.1.4 Two-part Tariffs

The costs incurred by an irrigation agency are largely fixed and usually dominated by annual maintenance costs and staff costs. Variable costs are usually only significant in areas where water is pumped.

A two-part tariff can meet an agency's need for assured income, while retaining incentives to farmers to avoid waste and use water productively. In this case the ISC includes a fixed charge (typically related to the area of the farm) with a variable charge based on the volume of water delivered, or a proxy such as area irrigated or the number of irrigation turns taken.

For a given level of total revenue a two-part ISC always results in a lower marginal price of water than in the case where charges are entirely volume-based. This is illustrated in Figure 3 where the gradient of the line indicates the marginal price paid. The reduction in marginal price reduces the incentive effect of price on demand, but where an agency prefers the revenue stability derived from a two-part tariff; this reduction in the *marginal* price paid by the user is an acceptable trade-off.

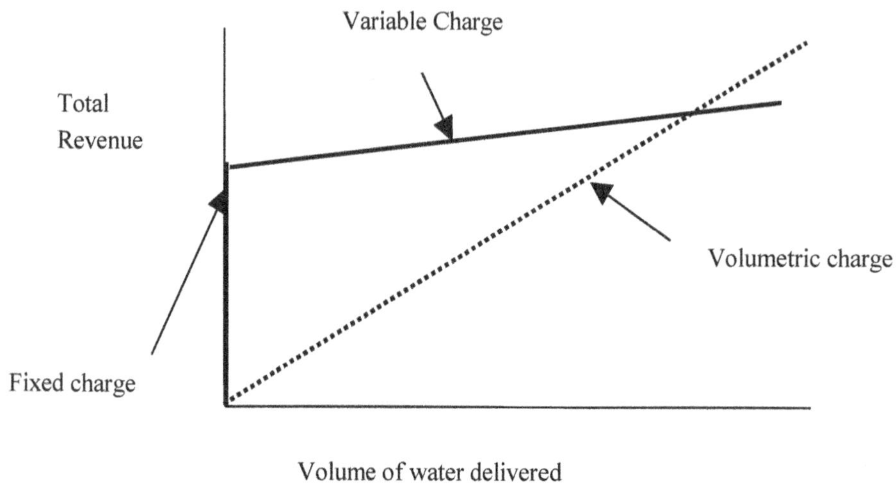

Figure 3 Volume-Based and Two-Part Irrigation Charge

4.1.5 Quotas

Rationing or quotas[4] (system 4) are widely practised in the field. Under this system, the farmer receives less than he could productively use, and he *perceives* its value in terms of crop output and income foregone if he wastes some of the quota. To the farmer this is the *opportunity cost* [5] of water. Where water is scarce and rationed, as in Haryana and Morocco (see Box 4.1), farmers use water carefully even though the marginal price to them is zero (in the case of Haryana) or well below its productive value (in Morocco). Where water is scarce and rationed, a farmer will generally take his allocated share of water (the marginal cost is low or zero, and the marginal value is high). Thus, a quota will limit consumption at a threshold but they will not provide any incentive to reduce water use below this limit.

> Quotas can be applied to meet any specified overall level of usage, and can be arranged to assign usage in ways that may be socially desirable (for example, by giving higher quotas or higher priority quotas to certain uses in conditions of scarcity, or assigning a quota to an environmental use).

A particular benefit of quotas is that farmers realise the value of water when their supplies are restricted. This may lead those formerly growing a crop with a high value of production per hectare, like sugar cane, to switch to a crop with a higher value of production per unit of water consumed - a desirable outcome in a water-scarce situation. To this extent, individual usage will be reduced in an efficient way, but at a wider scale there is no transfer of use from the less productive to more productive users or sectors. At a wider scale, quotas are transparent, and allow an open political debate about the volume assigned to particular uses, and priorities among users.

---

[4] Quotas and rationing are the same in that the physical volume assigned to a specific use or user is specified and does not vary with demand. The quantity may be either fixed and prioritised – so that the first priority user takes the assigned quota, followed by the second priority user, until no water is left and remaining quotas are unfilled, or may be based on a specified share of whatever water is available.

[5] More generally, the opportunity cost of water is what could have been produced in the best alternative use – perhaps in another sector altogether. Such costs are often apparently very high (for example if there are shortages of water for drinking while low-value crops are irrigated). Caution must be used in interpreting such values, because diversion of enough water – and often it is not much – to meet the full demands of the other sectors results in the opportunity cost rapidly falling back to the much lower, agricultural level.

Box 4.1 Not all quota systems require volumetric measurement

One of the most widely described examples of quotas is the warabandi system of northwest India and Pakistan - practised to some degree over some fifteen million hectares. The system combines shares and priorites; each hectare of land is entitled to an equal quantity of water in each irrigation turn, defined as a period of time during which irrigator may take water from the watercourse. Peer pressure works very effectively to enforce discipline among farmers. The periods during which watercourses run are based on a set of priorities, which rotate week by week. During high priority rotations a watercourse will almost certainly run and farmers take their quotas; during low priority rotations, a watercourse may get no water and a turn is missed. While easy and cheap to operate, the warabandi system relies heavily on the specific design of the infrastructure - and many attempts to introduce warabandi without the necessary infrastructure have failed.

A far more flexible quota system is operated in Morocco, where farmers are assigned a "water cheque book" entitling them to a defined volume of water for the season, according to a schedule that the farmer can largely determine.

While the warabandi system has very few operators and very few control structures, the Moroccan system requires highly sophisticated infrastructure and intensive management - resulting in O&M costs some hundred times higher, yet still affordable.

Administering quota-based systems

Quota systems may specify volume or share per season, per hectare, or per farmer. However, such rights are often qualified in some way. For example, the specified volume may be for an average hydrological year with reduced entitlements in low rainfall years or the right may be subject to priorities depending on the type of crop being irrigated, or the status of the user with respect to other users.

Any quota-based system carries with it significant administrative responsibilities for the operating agency. Examples of procedures for allocating water by means of a quota or defined share, include:

1. Abstraction rights, where the annual or seasonal volume that can be abstracted is defined without qualification.

2. Abstraction rights where the annual or seasonal volume is specified, but qualified with respect to actual water availability or other factors such as crop type.

3. Priority rights, where the sequence in which users (or sectors) are entitled to specified volumes of water is defined, and water is allocated on that basis until there is none left. This procedure is followed in the western US, where earlier rights have "seniority" whether upstream or downstream of "junior" rights.

4. A "sufficiency" basis for allocation of water in field-to-field rice irrigation in Asia. This is similar to 3, above, except that "seniority" is determined entirely by location, with upper users fully irrigating their fields before allowing water to pass to lower users - who in consequence have a lower assurance of supply. While this system is not a "quota" in the sense of a defined physical volume, it does provide rules defining priorities and a test of whether the quota has been met for each user.

5. Crop-based systems, where a particular cropping pattern is authorised seasonally for each farm, and the farmer is entitled to take "adequate" water to meet the needs of that crop.

6. Proportional rights, where the water available to the irrigation scheme is divided among users in accordance with their land holding.

7. Tradable rights, where its owner can sell a right based on one of the procedures above. In such systems, careful attention must be paid to third-party impacts (for example, recharge to aquifers, or return flows to rivers for navigational use). As a result, the quantity of water received by the buyer may be less than the quantity sold, with the balance being used to protect third parties.

These quite different procedures for allocation vary widely in terms of rationale for allocation, equity objectives and the means of confirming compliance. However, in each case the following requirements are met:

- Allocation procedures are defined
- They can be related to current water availability to establish who should be getting how much water
- Field observations can compare the assessed water allocation to actual deliveries.

Quotas and the need for volumetric measurement

Any of the procedures listed could provide a clear basis for specifying the agreed service and formulating a related charging system, but they require quite different abilities to measure water delivery, creating another precondition – if the chosen water allocation procedure depends on measurement, then the infrastructure must provide that possibility:

- In field-to-field systems (4), no measurements of water delivery are needed, nor are they feasible.
- In crop-based systems (5) no volume is defined or measured; the area of each crop and, from the total of the crop areas, the overall area irrigated, are measured as indicators that the agreed irrigation service has been provided.
- In proportional systems (6) the entitlement is essentially embodied in the design of the system and, provided the system is operated according to design, proportionality is delivered.

In systems 1, 2, 3, and 7 some form of volumetric measurement or assessment is implied, to ensure that the rules have been followed and the entitlement has been delivered. In some cases this is done at the farm level, but often the volumetric measurement is carried out at a higher level. Below that point either the system operator is entrusted with allocating the measured quantity among users, or indicators of volume, such as duration of delivery at an estimated flow rate, are used.

These systems are often referred to as volumetric, or providing "volumetric allocation" which, in turn, suggests attributes of a market, where a certain amount is ordered, supplied and paid for on the basis of an agreed price. However, there is a clear distinction between measurement to check that an entitlement has been met, and a system of volumetric delivery where the user is free to decide how much water will be taken on the basis of price considerations. In the former case, volume is measured to confirm that a set of rules, which define the water entitlement, have been complied with. A volumetric price may then be applied to the quantity delivered and, if it turns out that too little was delivered, the farmers pay less. However, they do NOT then have an option to pay extra and get the rest, nor can they demand more than the agreed quota on the basis that they are willing to pay more. In this context the "volumetric charge" is only a reflection of what was previously agreed, not a component of current decision-making.

> ***There is a clear distinction between measurement to check that an entitlement has been met, and a system of volumetric delivery where the user is free to decide how much water will be taken on the basis of price considerations.***

Systems 4 and 5 are open to misuse and dispute, because the test of whether the quota has been met is based on sufficiency of supply – which in turn depends on such factors as good field preparation. All the other systems provide an incentive for good on-farm water management.

4.1.6 Tradable water rights

Tradable water rights combine the benefits of quotas – assuring a match between supply and demand – with the additional potential to transfer water between users and indeed sectors. However, the system is not yet widely used due to the reasons that are set out in this section.

From the water users' perspective, tradable water rights offer three options:

- To continue irrigating, and pay the associated charge for the irrigation service
- To buy additional water in the market and expand irrigation activities
- To sell water and reduce or abandon irrigation.

Through these market transactions - where payments are entirely unrelated to the payment for irrigation services, water is transferred from lower to higher value uses - fully meeting the objective of increasing the productivity of water. Prices are market-determined, and the only intervention in the market by the regulatory authority is to address possible third-party impacts.

Tradable water rights are not a form of irrigation service charge but are the most effective way of meeting economic efficiency objectives, as well as ensuring the demand is limited to the available supply. However, the system requires a well-developed administration to control and monitor rights, and is poorly suited to situations where many, small entitlements exist.

Tradable water rights are not a form of irrigation service charge but are the most effective way of meeting economic efficiency objectives, as well as ensuring the demand is limited to the available supply. However, the system requires a well-developed administration to control and monitor rights, and is poorly suited to situations where many, small entitlements exist. A particular advantage, when compared to volumetric charging as an allocation mechanism, is that the market price of water is applicable only to actual sales: individual users who wish to continue to exercise their rights to use pay only the charges related to O&M and other agreed fees, not the much higher price that water commands in the market. But the fact that the "market" price is known to all users encourages rational decision-making.

Box 4.2 Tradable water rights in the Murray Darling Basin

In the Murray Darling Basin, water supplies are scarce and extremely erratic and soil salinity is severe. In a process that began more than 20 years ago, water rights, essentially based on historic patterns of use, have been formalised so that users have an entitlement (or entitlements) specified in terms of volume and security. Highly secure rights are met, or exceeded, in almost every year; less secure entitlements may only be met in 25% of years - so the water allocation rules are consistent with the erratic nature of water availability. Salinity entitlements - the rights of an area to *export* salt - are also specified, and each area must stay within its entitlement, or face significant financial penalties. Water deliveries are measured at the farm gate, primarily in order to confirm that entitlements have been taken.

Once water rights had been firmly established and documented, the possibility existed to allow trading in water rights. The system has developed in complexity over time. It now provides the possibility to buy and sell entitlements on a seasonal or permanent basis with a high or low security of entitlement.

It has recently become possible to trade water at significant distances that is, not within a local jurisdiction. This involves third party impacts on river flows, recharge, etc. Such inter-state trading involves a number of key components:

- Water "equivalence" ratios, which define, at the basin scale, what a unit of water in one place equates to in terms of water at another location. Thus, a purchaser of 100 units in location A may have to buy 120 units from Location B, or 95 units from Location C in order to have the supply he needs. (These ratios lead to variations in price between locations for a given quantity of water).
- Procedures within each State's water licensing authority to authorise sales and purchases (both states having to concur)
- Annual adjustments to State allocations by the Murray Darling Basin Commission, reflecting transfers.

The entire process of transferring a water right involves 12 steps. So far, water trading has been limited. Permanent transfers are going on at an average rate of about 1% of total availability each year. Temporary transfers are about 10% per year.

Tradable water rights provide the basis for re-allocation of water from lower value uses towards high value uses. Thus towns can be expected to purchase water rights from agricultural users without political judgements being required as to how much water domestic users should receive. Further, society can make judgements about appropriate allocations to environmental uses - and pay for the transfer to such use.

An important complication in the operation of tradable water rights is the need to monitor and react to third party impacts. For example, if a navigation use was made possible by water deliveries to a particular area, or the seepage from canals fed domestic wells, then a transfer of irrigation water use that affected these other uses would have to address the rights of such third party users.

4.1.7 Quotas, tradable water rights and revenue

Quotas are not a charging mechanism and will not generate revenue to cover the costs. If quotas are chosen as the means to limit demand, there are important implications for the charging system. If a volumetric system of charging is under consideration, using a quota will achieve the "demand management" function as well as providing the incentive to increase productivity. The extra benefit that volumetric measurement provides is clarity that the promised service has been provided, and the potential for more flexible scheduling among farmers while still keeping records of entitlements (compare the systems of rationing in Haryana and Morocco in Box 4.1) but that benefit must be set against the cost of volumetric measurement and charging.

Where a quota system is in force (whether tradable or not) a separate charging system is needed to ensure cost recovery. Where tradable water rights are used, it is important to distinguish between the payments that irrigators make to the agency for the service provided, and the separate payments made (or received) in the event that water is traded. The first payment is an ISC in the context of these Guidelines; the second is a capital payment (if the rights is bought in perpetuity) or a rental payment if the rights is bought for a specified period of time. In either case an administrative payment may be made to the organisation overseeing the trading process.

Where water is traded within or between projects and sectors, the potential revenue base for the operating agency changes. Indeed the size of the revenue base may vary much more sharply than the O&M costs which are mainly fixed. Projects where water is bought in will have a larger base over which to spread fixed operating costs, but projects selling water may have to increase charges for the remaining water in order to maintain the same level of income.

4.2 COMPATIBILITY OF OBJECTIVES AND TOOLS IN ISC

Table 4.2 summarises which objectives can be realised with the different charging systems and allocation procedures, taking account of the likely outcomes or behaviours that have been described in Sections 4.1.

Table 4.2 Compatibility of objectives, charging systems and other mechanisms

Objective	**Charging system or other mechanism**				
	Area-based	Crop-based	Volumetric	Quotas	Tradable water rights
Cost recovery	Yes	Yes	Possibly	No	No
Demand management	No	No	Possibly	Yes	Yes
Productivity	No	Yes	Yes	Yes	Yes

4.3 PRECONDITIONS FOR IMPLEMENTATION OF IRRIGATION SERVICE CHARGE SYSTEMS

Implementation of any charging system requires that wider political, legal, and administrative preconditions are satisfied. These require a major and on-going commitment of resources and political will and must not be under-estimated.

Where volumetric charging systems, tradable water rights or the use of quotas are considered, these carry additional implications for the irrigation infrastructure which is necessary to control and measure discharges to individual users or user groups. It is vital to recognise that the method of irrigation charging must be consistent with the design of the irrigation system and the method of irrigation scheduling and delivery that is employed. In particular, measurement of actual deliveries to small farmers is generally difficult – and impossible in field-to-field rice systems.

Physical modernisation of such schemes is possible, but the costs and technical demands implied by a change from continuous flow or rotational delivery to a system of variable delivery, according to demand, would be enormous.

Table 4.3 provides a summary of the legal, administrative and technical preconditions that must be in place for an irrigation charging system to stand a chance of working effectively. Some of these conditions are examined in more detail in the sections of these Guidelines considering the implementation of a charging policy.

Table 4.3 Preconditions for effective irrigation charging

Aspect	Detail
Legal	Users have legally defined and enforceable - and enforced - water entitlements and a system of allocation.
	A clear and viable judicial and police system exists to ensure enforcement of agreements.
	Corrupt practice is the exception and is acted against rather than being the tacitly accepted norm,
Administration	A clearly understood and agreed-upon charging and fee collection system, to include: 1. How fees are computed. 2. Mechanism for fee payment (to whom, and how) 3. How the fees are requested. 4. When fees are to be paid. 5. Penalties for non-payment of fees, or late payment of fees. 6. Whether the user(s) can refuse payment for water that was delivered, but not requested.
	A specified mechanism to resolve disputes over deliveries or bills.
	Sufficient human, technical and financial resources are provided to implement assessment, billing and fee collection.
Specific to volumetric charging	
Infrastructure	Infrastructure permits control and measurement of volumes delivered to users or a user group and means exist for users to verify volumes.
	Infrastructure permits delivery of differential volumes to neighbouring users.
Administration	A written agreement between the water supplier and the user(s) of the nature of water delivery service, to include: 1. Advance time required to order, change, or stop flow. 2. Other details related to the flexibility of frequency, rate, and duration of water delivery service. 3. Accuracy of the flow rate measurement device. 4. Allowable percent variation in the actual flow rate from the agreed-upon flow rate at any time. 5. Who can make the flow rate changes (the supplier or user) at the control structure. 6. How frequently the flow rate can be changed. 7. How frequently the flow rate must be verified, and how. 8. Responsibility for maintenance of the measurement and control structures. 9. Penalties to the water supplier if structures are not maintained or operated as specified, or if the quality of water delivery service is poorer than agreed upon. 10. A procedure for when, and how, any volumetric limitations are determined.

Modified from Burt (2002)

4.4 CHAPTER SUMMARY

See Table 4.4.

Table 4.4 The basis for irrigation charging – chapter summary

1. Understand the effects of charging systems and other economic tools:

Type	Description	Impact on productivity	Impact on demand	Can assure supply/demand balance?	Stability and predictability of revenue	Ease of administration
Area-based	Fixed according to farm area	None	None	No	Good	Good
	Fixed charge per area irrigated	None	Small	No	Moderate	Good
Crop-based	Rate based on crop type irrigated	Small (a)	Small	No	Moderate	Moderate
Volumetric	Fixed rate per unit volume received	Positive (b)	Positive	No	Poor	Poor
	Rising block tariff	Positive	Positive	No	Poor	Poor
Quota	Entitlement to water is defined	Positive	Controlling	Yes	Not relevant	Variable
Tradable water rights	Entitlement to water is defined and can be traded	High (c)	Controlling (d)	Yes	Not relevant	Low

(a) Small = essentially no useful impact except at extremely high levels of charging
(b) Positive = Will promote movement in the desired direction with magnitude dependent on level of charge
(c) High = Impact substantially independent of the charging system
(d) Controlling = Specifies the maximum level of demand that will be satisfied under different supply conditions

A two-part tariff combines a fixed charge per irrigable hectare plus a variable charge, based on volume used, number of irrigations or irrigated area. The fixed element provides stability of income whilst the variable provides incentive to reduce waste and improve productivity.

Rising block tariffs these provide strong incentives to keep consumption below the threshold where high tariffs apply. However, they rely upon accurate volumetric measurement of supply to users. Present use in the irrigation sector appears limited to Israel and Jordan where water is supplied to users via pressurised pipe networks

Quotas and tradable water rights allow the control of demand but do not generate revenue. For practical application in many developing countries a quota may need to be defined for users groups and not for individual users. Most systems of allocation by quota will require volumetric measurement to confirm the quota entitlement has been delivered.

2. Select the charging type that best meets the primary objectives and that can be applied in the field, given the availability of staff, methods of water management and system infrastructure.

3. Check that key preconditions concerning legislation, administration, and infrastructure are satisfied – see Table 4.3

5. Quantifying the charges

Previous chapters have dealt with the policies and objectives which need to be clearly defined before contemplating the scale, scope and application of the charge to be set.

This chapter deals with issues involved in setting an appropriate charge:

- the costs to be recovered
- how much to charge
- who sets and who updates the charge
- allocating the charge to users
- special provisions, discounts, rebates, compensation

5.1 HOW MUCH TO CHARGE

5.1.1 Costs of service: Introduction

Charges should preferably be linked directly and transparently to costs, but they should not be set unnecessarily high so as to cover waste and mismanagement by the irrigation provider.

There is a risk that political pressure may cause irrigation charges to be reduced below the level which is desirable. It is therefore important that charges are decided by a well-respected board (Section 5.2.1). Ideally, politicians should agree to the methodology of calculating the charges and phasing them in, rather than have involvement in the details of setting the rates. Oversight should be provided by a regulatory or auditing committee involving stakeholders, represented by at least the irrigation provider and water users.

Charges need to be updated regularly according to agreed procedures, principally to allow for the effects of inflation. In practice, charges are often not updated, meaning that they fall ever further behind operational expenditures. Corrections can be made on the basis of inflation rates or of the actual expenditures incurred. Where irrigation charges are seen as taxes that do not directly contribute to the financial resources of the irrigation provider, a correction is typically made on the basis of the inflation rates recorded in national economic statistics. Where they are seen as service charges to an autonomous irrigation provider, the correction is best made on the basis of the business plan, taking into account the cost increases in the various activities and possible tasks added or deleted from the overall programme.

If inflation is high, correction is difficult. As an example, in Turkey, during a period of high inflation, charges were based on actual expenditures made two years earlier. As a result, there were huge shortfalls in funding. A system of partial pre-payment is more appropriate in such circumstances.

When irrigation charges are based on real costs, two broad categories of cost may be considered:

- operation and maintenance costs
- capital costs

A third category, resource or opportunity costs (usually based on the economic value of irrigation water in alternative uses), is sometimes also considered. Such costs are rarely taken into account in practice, however, in large part because of major uncertainties in practical assessment.

The costs calculated in each category can be factored by a target percentage recovery figure and aggregated to give a generalized irrigation charge. When it is aimed to recover anything less than 100% of the costs, it must be clear how the difference is to be paid for. Similarly, it should be decided how a failure to recover dues is to be dealt with. For example, in Moroccan water management organizations where pumping accounts for a substantial part of the operational costs and collections are less than 100%, there is an annual transfer from the central budget to fund part of the shortfall and to meet capital costs. On the other hand, it may be decided to recover more than 100% of costs, to create a buffer fund for contingencies and disasters.

The calculation of real costs may be complicated by the fact that the financial records of irrigation providers make it difficult to trace expenditures. Typical problems are:

- Separate cost centres may not be defined, making it difficult to isolate expenditures. This is particularly problematic if the irrigation provider is involved in a range of activities besides irrigation, as in many of the systems in Eastern Europe which also operate construction companies, hotels and other businesses.
- Expenditures on irrigation management may be made by several organizations without anyone knowing the total amount that is actually spent.
- Expenditures on system maintenance may be itemized under a series of budgetary headings apart from maintenance, such as Improvements, Emergency Works, Establishment Costs (salaries) etc.
- The total expenditure on maintenance is frequently inadequate to sustain the system in good condition, resulting in a general decline.

The financial administration of the irrigation system needs to provide an accurate picture of actual expenditures, in order to review the basis for irrigation charges and to allow adequate control of costs.

Box 5.1 Tracking irrigation expenditures

On the Nara Canal in Sindh, Pakistan 65% of the expenditures on the system were incurred on 10% of the command area - a series of 15 upstream canals supplied by pumping stations. Because these costs were paid directly by the Provincial Government to the Federal Electricity Board, the irrigation agency had no real understanding of the magnitude of overall expenditure.

Source: Euroconsult, 1998

It is also important to keep down the costs. Realistic costs to provide effective services should be charged, preferably based on best practices, norms and standards in a country or region. Costs resulting from over-staffing, inefficiency and corruption must be removed before the costs are passed on to users.

Irrigation providers need to take other measures to improve their financial performance and ensure that their primary function of water delivery is assured. Other sources of finance can be sought and services rationalised. Important savings or additional revenues are possible when irrigation agencies decide to operate in a more entrepreneurial and service-oriented manner. It is advisable to calculate charges on the basis of a business plan rather than on an expenditure sheet. A wide range of measures to balance income and expenditures should be explored in the plan - not just by adjustment to the water charges, but also by rationalizing costs and developing additional sources of water-related income. Typical contents of such business plans are given in Table 5.2.

Table 5.2 Elements in an irrigation provider's business plan

1. Setting the right charges
2. Improving collection efficiency
3. Rationalizing costs
4. Rationalizing scope of services, including
   - Dispose of non-core activities
   - Close loss making sub-systems
   - Transfer management responsibilities to users
5. Exploiting assets and resources
   - Buildings
   - Land
   - Fishery concessions
   - Timber concessions
6. Extend service to industrial, municipal and other consumers

5.1.2 Calculating Operation and Maintenance Costs

The O&M budget for a given scheme may be estimated by a number of methods:

- Historic costs
- Best practice norms and standards
- Needs-based assessments
- Performance contracts

Historic costs

There are two major problems with using historic costs. First, O&M on many irrigation systems has been chronically under-funded, so current or historic expenditures do not reflect the level of expenditure required to sustain a satisfactory level of performance. Budget requests are often drawn up to support a bargaining process and may not be directly related to actual requirements. Secondly, in countries suffering from high rates of inflation, even current or very recent O&M expenditures will not adequately reflect the amount which needs to be recovered from farmers in the coming season to maintain the real level of expenditure. In these circumstances, realistic allowance must be made for the likely effects of inflation in eroding value, when determining the level of the charge.

It may be possible to use historic maintenance norms and standards. Where most of the operation and maintenance costs are predictable, this method may work well, provided due allowance is made for inflation. In many cases, however, the method is inflexible and special needs can differ from year to year.

Best Practice Norms and Standards

Recognising that operation and maintenance are consistently under-funded, many countries have aimed to identify target levels of funding to keep systems in good condition. However, there are many variables which affect the needs for O&M e.g. system age; water source (surface, reservoir-backed, groundwater); system design characteristics; prevailing soil, topography and climatic conditions. Thus, whilst such norms provide a useful guide, it must be recognised that they may not be directly applicable to individual cases.

Needs-based assessment

> ***Needs-based budgets potentially provide the best indication of real costs, and provide a mechanism to account for inflation.***

Needs-based budgets potentially provide the best indication of real costs, and can allow for inflation. However, they require a formalised set of procedures, to be applied year-in and year-out ("asset management") and a systematic description of key operation and maintenance activities, including a breakdown of the resources required - labour, material and equipment. The most important operation and maintenance costs on a given scheme will include many of the following:

- *Operation:* data collection, water measurement, operation of structures - including flushing and flood protection, inspection, preparation and implementation of water distribution plan.
- *Routine maintenance:* cleaning, painting, weed control, safety measures.
- *Periodic maintenance:* repairs, desilting, restoration of embankments.
- *Overheads:* salaries and wages, support staff and administration, elections, inspections and registration, taxes, insurance fees, compensation payments.

Replacement costs, at the end of component design life, are covered under capital costs in Section 5.1.3.

For each of these four components, unit costs need to be adjusted for increases in labour, material or equipment costs. Maintenance requirements for the year ahead (and also replacement costs) are assessed on the basis of a status survey, an inventory of the condition of the system. Field assessors need to be trained to follow a consistent procedure. A good system of record keeping - preferably a computerized database - is necessary to record historic expenditures, by category and location, and to identify changes of system condition over time.

Once needs have been determined, priorities will need to be set to make budgets fit within reasonable spending levels, which will certainly be constrained by previous years' budgets.

Box 5.2 National allocation of maintenance budgets

In Turkey, the Regional Directorates of DSI estimate their O&M requirements on the basis of annual inspection reports and standardized unit rates. The regional estimates are reviewed and revised by the Budgetary and O&M Departments of the General Directorate. Adjustments are made in the light of the available budget, using judgements as to which maintenance works can be deferred. The Directorate then decides on the regional distribution of the funding. Cuts are distributed proportionally over the regions.

Source:Scheumann, 1997

Many systems suffer from deferred maintenance. A backlog of work will need to be cleared before routine maintenance is capable of stemming further decline. Sometimes, a rehabilitation project may be carried out to address the problem. Most importantly for setting irrigation charges, a charge based on immediate needs would be much higher than commonly considered for routine maintenance and it might well exceed what farmers were willing to pay.

Many irrigation systems are in effect 'living infrastructure' where change, particularly in the command area, is an almost permanent feature. New structures and canals may be added, whereas other parts of a system may be selectively neglected because they are no longer needed or are too difficult and costly to keep operational. It is necessary to assess at regular intervals which infrastructure needs to be maintained and which can be decommissioned.

Maintenance should aim to keep the system functioning efficiently in a sustainable fashion. It is uneconomic and inefficient to unthinkingly restore all elements of infrastructure to their originally-constructed state, since conditions and needs may well have changed in the meanwhile. In practice, trade-offs will have to be made and the functionality of the different elements will need to be frequently reassessed for their fitness for function. The O&M budget for different regions will need to be based on locally-prevailing unit costs, which may even vary between systems. In practice, it may be useful to develop cost "ranges", so that each individual system is not greatly different from its neighbours. It is useful to compare between schemes and to set target "benchmarks", as it helps to better justify unit costs and identify areas of potential cost savings.

A simplified methodology could be suitable for farmer managed irrigation schemes, but in the initial stages, farmers are likely to need help from technical personnel to determine routine maintenance needs and to set budgets. There may otherwise be a tendency to postpone routine works until a serious situation has developed requiring major expenditure. Traditional irrigation providers in various countries are adapting themselves to new roles in supporting farmers' associations. Continuing technical assistance at an advisory level is a particularly valuable function for a restructured irrigation provider.

Performance Contracts

Under a performance contract, an operator will provide services against a negotiated sum which then becomes the O&M budget. The services are described in performance indicators. Whereas in the domestic water supply sector such contracts have become common, there are as yet very few examples of performance contracts in irrigation and drainage.

In Sindh, Pakistan, for the maintenance of the Left Bank Outfall Drain, it was decided to let performance contracts to private contractors, to avoid the creation of a large public body. In the initial year, contractors were cautious about bidding against an unknown quantity - their lack of familiarity led to inflated prices. Later, prices fell as experience was gained. The quality of work was variable- maintenance of minor facilities tended to be poor, since there was little regular inspection.

One risk with performance contracts is that it is easy for the contractor to reduce the scope of work when the budget is limited. Where maintenance is done in-house, execution may be less efficient but the investment in staff and equipment tends to ensure that a basic amount of work is done, though it may not be well targeted.

5.1.3 Calculating capital costs

The term full cost recovery generally implies the recovery of capital costs in addition to operation and maintenance costs, although in some contexts a wider set of costs is also considered. The common definition is assumed here. Public sector irrigation capital costs are rarely recovered in full, as the examples in Box 5.6 show, the shortfall in recovery being a deliberate government subsidy to irrigation users.

Box 5.3 Examples of Countries/States which subsidise capital costs

- The Chilean government meets 45% of capital costs, so only 55% needs to be recovered in charges.
- In Morocco, according to the 1969 Code, O&M cost are to be fully recovered, as well as 40% of the capital costs.
- In Mexico and Philippines, the cost of large capital projects is borne by the state.
- In Egypt, farmers repay the costs of sub-surface drainage over 20 years in interest-free annual instalments, with a grace period of 3 years after construction. (effectively, land owners pay less than 60% of costs, at an annual inflation rate of 5%)
- In Macedonia, depreciation costs were supposed to be recovered in the water price, but in reality this did not happen.
- In India, (and numbers of other countries), capital costs are not charged to water users, even though the Committee on Pricing of Irrigation Water recommended that charges should cover O&M costs plus 1% interest on capital investment).

For the purposes of irrigation charging, capital costs should be divided into two categories:

- **'Replacement'** capital costs: items such as pumps, vehicles and other machinery and equipment, which have to be replaced at regular intervals, and for which adequate arrangements for funding replacement need to be made in an irrigation system's financial planning and charging system.

- **'Non-replacement'** capital costs: one-off capital expenditures incurred at the scheme development stage, such as land acquisition and resettlement, and major infrastructure like headworks and large canals which require replacement rarely, if at all.

Each type of capital cost requires a different treatment in charging.

Replacement capital costs

To achieve full financial sustainability, so that, once built, a scheme does not require outside funding in order to continue to operate at full performance, all replacement capital costs should be recovered. If they are not, outside funding will be needed at intervals, usually from government, or replacements will have to be deferred, leading to decline in scheme performance. In practice, full replacement cost recovery is not achieved on government irrigation schemes in most developing countries, and in many, the level of government funding of replacement expenditure has been insufficient to make up the shortfall.

For calculating irrigation charges, there are two methods to determine the amount which should be included to cover replacement costs:

- *The Depreciation Method*, which is the most commonly used
- *Renewal accounting*

The *Depreciation Method* involves setting aside a sum each year, which is sufficient to cover future replacement expenditures, taking account of the expected useful working lives of all the assets to be replaced. In the past, depreciation allowances have normally been based on the historic costs of the assets, but a better method is to base them on current costs (the Current Cost Accounting system); i.e. what it would cost to replace the assets now.

Depreciation allowances in turn can be calculated on two bases; either taking account of the interest that can be earned on the sums set aside (a sinking fund), or ignoring it: this latter can be termed the simple depreciation method. A sinking fund can be defined as a fund which anticipates the renewal of a capital asset by setting aside a given sum each year which, including the interest earned on it, will provide sufficient money to replace the capital asset at the end of its useful life.

Either of these two depreciation methods can be used in calculating irrigation charges, depending on local practice and policies in the country concerned. The important thing is that adequate allowance should be made for replacement costs – the choice of method to do so is less important. On most developing country schemes, however, replacement costs are not recovered in the irrigation charges made.

Renewal accounting is an alternative to using actual capital costs or depreciation methods. In renewal accounting, the central concern is the long-term cost involved in renewal and rehabilitation of the system, rather than the original value of the system.

Box 5.4 Renewal accounting, Australia

Before 1984, water prices in Victoria State, Australia were not calculated on the basis of a systematic assessment of costs. However, a new government specified a number of financial objectives for the River Water Commission: increased levels of cost recovery, encouragement of profitable irrigation enterprises, and pricing signals on the value of water to its users. The Rural Water Commission was to recover all operation, maintenance and administration costs, as well as capital costs over a 20-year period. It was decided to take a 0% opportunity costs of capital (in effect, a real interest rate of zero), as any other policy would have increased the water price unacceptably. A system of current cost depreciation, as above, was introduced. To achieve the twenty-year cost recovery target, a substantial increase in water price, 11%, still had to be made. It triggered a vehement 'rate protest', in which farmers withheld payment. This resulted in a review of the pricing policy. Following the review, current cost accounting was replaced by renewal accounting, which made it possible to make a more realistic and modest estimate of long-term replacement requirements.

Source:Langford et al., 1997.

Renewal accounting is linked to formal methods of asset management. It requires estimation of the timing and cost of replacing infrastructure, after considering the risks and consequences of failure of key infrastructure. The condition of all assets is regularly assessed and rated during field inspections. A risk rating is assigned to each asset, taking account of the time it would take to reinstate the asset and the lives/property/production potentially at risk if failure were to occur. The cost of replacing the different assets is estimated, to determine the cash flow required to fund replacement. The procedures are more labour-intensive and costly in the short-term than other methods, but allow a more accurate assessment of the long-term replacement costs. In Victoria State in Australia, the estimated capital cost under renewal accounting was 25% less than the corresponding amount under current cost accounting, because replacement costs were less than the original investment costs.

Non-replacement Capital Costs

Recovery of non-replacement capital costs is of no direct relevance to scheme sustainability, because it does not affect maintenance of a scheme's asset base and performance levels. Where non-replacement costs are to be recovered, it is normally done through an *'amortisation'* charge. This is usually the most convenient way of charging for such costs.

Amortisation is the process of spreading the cost of an asset over a specified number of years (the 'amortisation period'), with or without interest. Where interest is charged, the amortisation amount is the annual sum of money which it will be necessary to set aside so that capital and interest costs are paid in full within the given amortisation period.

Of the three irrigation cost elements - O&M, replacement capital and non-replacement capital - recovery of non-replacement capital costs should be given the lowest priority. It should be the last of the three types of cost to be recovered because it is of no direct consequence to scheme sustainability. In most developing country schemes, no attempt is made to recover such costs.

Where amortisation charges are made, they should be, as far as possible, fixed on a basis such as unit of land or, at most, unit of irrigated land. Because non-replacement capital costs are entirely fixed, their recovery should not be included in any volumetric charging system.

5.1.4 Resource costs

The term *resource costs* can be interpreted in various possible ways. For the purposes of this document, a broad term linked to the value of the water resource and the resource system of which water is part, will be used. The issues involved are a matter of on-going debate, and in reality it is unusual to charge for resource costs. Brief notes are included here, principally to indicate the complexity of the issues, rather than to advocate inclusion of such costs.

Resource costs can be based on one or more of the following:

- Opportunity cost: the value of other water uses foregone owing to the use of water in irrigation. The value in the next highest use determines the economic cost.
- Environmental costs – based on the value of the harm done to other sectors, including the intrinsic value of the ecosystem, when water is withheld or delivered under a different regime.
- The increased or decreased value of local resource systems – expressed as the combined price of the different resources.

The different calculations are often complex. Complexity comes from several sources, such as the need to value water not only in its primary use, but also in other uses; uncertainty about how to assign value to deterioration in water quality; uncertainty about the worth of a resource which derives its value when combined with other resources (land, capital, labour etc.); above all, general difficulty in valuing resource uses. Decisions are required on the timespan of analysis and the weighting given to different impacts, including impacts that have become irreversible.

A further complicating factor is change in value of local resource systems. It could be argued that any part of any increase should be recouped from those who have an entitlement to the water resource and who see their property (often land) increase in value.

Resource costing may principally have value in reviewing different water management strategies in river basin planning.

5.1.5 Developing other sources of revenue in irrigation systems

Irrigation and drainage systems serve a large number of actual and potential functions – other than agricultural water supply. This often represent a large source of potential income for the management of the irrigation systems. In Sindh in Pakistan for instance a study into 'non-abiana (water charge) revenue options' suggested that the current dismal financial performance in the canal systems could be set right by developing other sources of income than the low area-based water charges to farmers. One alternative strategy was to charge non-agricultural users for water supply and for drainage functions. At present municipal and industrial users were hardly taxed for the supply of water to their estates. In addition the study identified a large number of 'business opportunities': for instance property and tourism development near Lake Kinjar, an attractive lake at close distance to the Karachi metriopole or near some of the barrages and along the canals near to Karachi. Other business opportunities were tree planting concessions along the large stretches of canal and drain banks and developing fishery potential in the irrigation system. Also the Irrigation and Drainage Authority historically owned considerable land in city centres, which it had neglected and had allowed to be encroached. This again represented a considerable source of potential income. The same applied for its guesthouses, now often no longer used, but attractive property for local functions. The estimated return from these non abiana revenue sources were estimated to be be of a similar order of magnitude as the irrigator charges. Moreover, they were relatively easier to collect. Table 5.3 is an overview of the different functions of irrigation systems, that can be capitalized and monetized.

Table 5.3: List of Functions in Irrigation and Drainage: Agricultural water supply

- Controlling water table for agricultural production
- Improving land accessibility
- Improving soil chemistry
- Increased property values of land near water fronts
- Leisure opportunities - water based recreation, golf resorts
- Domestic water supply
- Industrial water supply
- Water supply to other users
- Improved protection against floods
- Use of canal and drain bank for tree cultivation
- Defense lines
- Use of canal and drain banks for transportation
- Buffering water stock
- Generating water for reuse
- Effluent disposal
- Washing functions
- Livestock water supply
- Fisheries
- Navigation and ferrying
- Improved public health
- Reduced damage to built up property by controlling soil moisture
- Reduced incidence of water borne vector diseases through environmental sanitation
- Firefighting resources

In general, improving and effectively operating irrigation and drainage in an area (or for that matter any other water infrastructure) tends to affect, usually positive, the value of these different functions. Water resource development for instance also makes it possible to facilitate the development of functions that are high on the 'value ladder' - in particular leisure, housing or industries. The values of these functions go to different groups of stakeholders. These may be general interests (for instance improved public health) or private interests (use of canal bank for cultivation).

In capturing the values there are three models to recoup the costs:

- General charges - taxes and public subsidies
- User charges - for specific user categories
- Controlled business development - in particular public private partnerships.

In case of general interests the function is best charged to the general functions. Examples are flood protection and the improvement of public health functions.

Where there are specific categories of 'users', charges should be put to those categories of users, as far as practical - with minor users possibly excluded so as to minimize transaction costs. Specific users apply in case of: irrigation, domestic water supply, industrial water supply, effluent disposal.

A third type of function needs business development to become operational and financially effective. These are: the development of water front property, canal bank forestry, leisure opportunities, commercial use of vegetation in water ways etc. In these cases public private partnerships (PPPs) may often be the best option. In these PPPs the challenge for the public sector is to ensure that value increases are not just captured by other parties, but they are rerouted to cover investment and running costs.

Box 5.5: Examples of improving non agricultural revenues examples

In the **Netherlands** the strategy to capitalize on these values increases is called 'red for blue' (use income from real estate - i.e. bricks - to pay for water investments and functions) or 'red for green' (paying for investments in environmental functions. An example of such an integrated project is the 'Nieuwe Hollandse Waterlinie', where an area of 20,000 ha is developed - with a range of functions: leisure, water management and housing development. The project was managed by board representing the regional authorities and the initiative was taken by private property developers. Of the 180 Million Euro cost of the project 40% was recouped from income from real estate. Similarly investment in flood protection is recouped by giving out attractive building plots on the reinforced embankments.

In the West Delta Project in **Egypt**, there are a number of high value functions that are now sustained by finite groundwater resources, but will soon be supplied by a surface canal. An example is the golf court and high value residential areas. These are high value functions and should contribute accordingly to the development and operation of the new West Delta canal.

In the **Senegal river** the opportunity to use a commercial private party to 'harvest' the excess weed growth in the water ways and convert it into 'briquettes' for export is being considered. This would turn waste into an asset and create value and employment opportunities in the process.

5.2 CHARGING WATER USERS

As stated in Section 5.1.1, it must be established whether different costs are to be fully met, partially met, or not met by the user. If costs are not fully recovered from the charges, an alternative source of funding needs to be defined.

A number of decisions need to be made:

- Who sets the charges and who updates them.
- Whether to set a standard single charge or varying charges.

- Whether to make special provisions for the occurrence of floods, droughts and other natural disasters which reduce farmers' incomes and capacity to pay.

5.2.1 Responsibility for setting/updating charges

The body responsible for setting charges will vary from country to country. However, it is considered that a procedure which includes the following actors is desirable:

- A central government sub-committee providing constitutional authority for setting a charge, responsible for issuing directives as to the objectives of charging and any relevant policies on partial cost recovery, subsidies and so on.
- A panel of experts, including representatives from the irrigation and water resources sector, to advise government and the regions on a suitable range of irrigation service charges allowing for differences in conditions across the country. The Vaidyanathan Committee in India is an example. The panel should convene at least annually to review recommendations in the light of inflation and the general cost index.
- State/local government committees, including WUA representation where relevant, to set, agree and annually review charges which are consistent with the recommendations of the national panel and appropriate to the schemes within their areas of jurisdiction.

The dangers of allowing elected politicians to set or decide on charges are too obvious and commonly observed, to warrant further emphasis.

5.2.2 Setting varying charges

It may be decided to allow differential charges, targeted to:

- Different areas/schemes.
- Different levels of service.

As indicated previously, many variables will affect the needs for O&M and therefore the charge required to recover costs: the system age; water source (surface, reservoir-backed, groundwater); system design characteristics; prevailing soil, topographic and climatic conditions, are just some.

Many countries have accepted the logic of charging differentially between areas or schemes. However, where radical differences in charges will result, it may be decided at national level (see Special Provisions, below) to cap charges on the most expensive schemes, in the interests of social welfare.

Where schemes are very large, it may seem logical to vary charges across the scheme. There could be several reasons to use different charges for different parts of a scheme. The quality of irrigation service can vary from one area to another (head-tail; water quality). The cost of supplying the water may differ greatly from one place to another; in particular, pumped supply is significantly more expensive than gravity supply. Similar considerations apply in drained areas. It is important to have a good understanding of the cost of services to different part of the irrigation system, so that an informed political decision can be made as to whether the entire excess costs are charged to local sub-units or whether cross subsidization is allowed between different parts of an irrigation system. In practice, the bureaucracy required to assess and administer such a system of varying charges could be prohibitive.

The irrigation provider may also introduce different rates to reflect different levels of service. Different rates are often linked to the point of supply; for example, at system head or at farm inlet level. Another practice is followed in Morocco, where water charges are uniform throughout the system, but rebates are given to farmers who undertake the cleaning of the tertiary channels (20% rebate) or both the tertiary and secondary channels (40%). In Macedonia, a 50% rebate was given to water users associations which took over the entire management of water at tertiary level.

Special provisions

The irrigation provider may consider giving rebates or compensation payments, for a number of reasons:

- To reward 'good' water management behaviour - such as the use of water-saving devices or cultivation which improves water productivity.
- To promote the use of off-peak water, by giving reductions for the use of night-time supplies.
- Relief at times of floods, droughts and other natural disasters which reduce farmers' incomes and payment capacity.
- To compensate for inadequate service. Such rebates are only possible where water is paid for on the basis of supply. Elsewhere, compensation is sometimes paid for damage resulting from leakage.

There is, however, a genuine risk that rebates will be given too freely, and that corruption will undermine the charging system. Another risk is that rebates will be given for administrative reasons, to disguise shortfalls in irrigation charge collection. Instead of relegating waived payments to a contingency allowance, a clear budget head should be established for rebates.

5.3 CHAPTER SUMMARY

See Table 5.4

Table 5.3 Quantifying the charges – chapter summary

Step	Actions/Issues	Recommendations / notes
Check broader issues	• Ensure national policy has defined which elements of cost to recover.	Seek clarification if necessary before progressing.
	• Identify fund/mechanism to cover costs not recovered from irrigation service charges .	Examine potential to recover costs by provision of water to higher value users – municipal and industry – whilst identifying consequences for agriculture
	• Identify opportunities to reduce costs	Review staff complements, as over-staffing of irrigation agencies commonly leads to high overhead costs.
Calculate O&M costs	ALTERNATES: 1) Needs-based budget OR 2) Best practice norms and standards OR 3) Historic costs	**Recommended method** = (1) provides best estimate of true costs and takes account of inflation, but needs more resources, good records and regular assessments. (2) Does not reflect differences between schemes (3) Not recommended. May reflect historic under-funding and inflation makes historic costs of meaningless.
Calculate the capital costs	• Separate replacement and non-replacement capital costs	Replacement = items known to need replacing at regular intervals e.g. vehicles, pumps other equipment. Non-replacement = major infrastructural works.
	• Determine if non-replacement capital costs are to be recovered from users.	Recovery of non-replacement capital costs is not necessary for the long term, financial sustainability of the scheme
	• Decide between depreciation and renewal accounting methods to determine the annual cost to be recovered. ALTERNATES: Depreciation based on 1) current replacement cost OR 2) historic costs OR Renewal accounting	**Recommended method** = (1) avoids complex evaluation of assets but budgets for replacement. (2) The historic cost is insufficient to cover future replacement. (3) More complex procedure based on assessment of asset condition and risk of failure.

Table 5.3 Quantifying the charges – chapter summary (continued)

Establish a body with authority to set and update charges.	Ensure the body has the statutory authority to act and is not linked to political interests.	
	Ensure adequate representation of stakeholders.	The make-up of the group will be influenced by the objectives of charging. If this is restricted to cost recovery it may not be necessary to include representatives of other water sectors.
	Ensure technical competence of the body.	Aim to limit inappropriate political influence on decisions.
	Specify the basis for, and timing of, regular review and updating of charges.	
Variation of charge rates	Agree at what level charges may vary to reflect different conditions: • No variation • Within a single scheme • Between schemes • Between administrative areas	Workable level of variation will depend on level of autonomy of different management groups plus any need to subsidise high cost areas from revenues from low cost areas.
	Define special provisions for the poor, natural disasters or crop failures etc. Consider rebates for early payment.	Keep to a minimum and ensure clear and transparent specification of the grounds for use.

6. Implementing an irrigation charging system

The previous sections have dealt with the considerations involved in structuring a viable charging system. These are a necessary, but insufficient, condition for setting up a successful system.

Strategies for water charging depend critically on the effective assessment and collection of water charges. An efficient charging system will achieve a high revenue performance with low transaction costs.

The present chapter addresses issues which are basic to establishing and sustaining effective revenue collection, namely: assessment and cross-checking of assessment, billing, collection at reasonable transaction cost; measures to improve level of collection, sanctions for non-payment; monitoring and feedback, so that management and an oversight committee are able to assess and improve the system.

6.1 ASSESSMENT

The scale of the assessment task, and the resources required for it, are determined by the type of charging system selected. Systems based on farm area, rather than crop, are relatively easy to assess, but they are rare in the developing world. Crop area-based charging systems, which are most common in developing countries - for example, the Shejpali system in India - demand substantial resources and time to obtain a meaningful assessment.

Under-assessment can be as big a problem as non-payment. Corruption can undermine willingness to pay, and lead to low recovery rates.

Decisions need to be made about:

- The categories of assessment
- Carrying out the assessment
- Checking the assessment

6.1.1 Categories of assessment

In general, the simpler the assessment system, the smaller the risk of abuse and the lower the transaction costs. If a large number of categories are established, each with different charges, the complexity and cost of assessment is increased. In Pakistan, more than fifteen categories of land use have been used to assess irrigation charges and many special arrangements existed – for example, for areas with drainage services, areas served by pumps and areas affected by disasters. The chance of misclassification, by accident or by manipulation, is correspondingly increased. In crop-based assessment systems, the number of different crop categories should be kept to a practical minimum.

> ***In general, the simpler the assessment system, the smaller the risk of abuse and the lower the transaction costs. If a large number of categories are established the chance of misclassification, by accident or by manipulation, is correspondingly increased.***

6.1.2 Carrying out assessment

Assessment may involve reading water meters, identifying crops in the field, measuring crop areas, and possibly, decisions about whom to exempt from payment. When there are many small water users, the costs involved in assessment can take up a significant part of the revenues, if not kept in check. Estimates from Macedonia put collection costs at 16% of the income from relatively high charges. In a study in Bangladesh, collection costs were estimated at 20% of the charges due.

The high cost of assessment is one reason to consider multi-level water charging, under which the main agency supplies water in bulk and the assessment in the sub-units is done by separate organizations. In water metered schemes for instance, if customers are delegated responsibility, they become responsible for tampering or wilful damage to the meters. If repair costs are automatically charged to the customer, there is a reasonable chance of preventing damage.

Preferably, the assessment process should be combined with other field activities. Whether the assessment is made by irrigation provider staff, by Revenue staff or by others, it is important to make cross-checks to ensure that water deliveries match those recorded and charged for.

If an outside organization is engaged, it is important that its mandate make it accountable to the irrigation provider. In Sindh, Pakistan and in Bangladesh, where the Revenue Department has traditionally collected the water charges along with other taxes, there has been frequent disagreement between the two departments about the level and collection of charges.

6.1.3 Checking the assessment

To help ensure that assessment is fair and reasonable, cross-checks should be included in the process. As a simple rule, if assessment is based on output, then the input should be cross-checked. Alternatively, if assessment is based on input (water delivered), the check can be on output - the area under cultivation. If such cross-checks are not possible, then as a minimum, the revenue from sub-areas should be compared, to identify good and poor performance.

Cross-checking of irrigated areas can be done most simply by visual inspection, or by transects and/or remote sensing. In a transect, an area is systematically traversed in a grid pattern. The land use is recorded at systematic intervals, thus building up a sample of the overall area. Transects are appropriate for smaller areas - two people can cover an area of 20 hectares in a day. If a larger area is to be covered, a multi-level transect can be considered, in which a larger sample is built up from the sub areas covered by a detailed transect. Satellite imagery can be considered for larger areas. In recent years, much work has been done in developing algorithms that allow land use to be classified and water stress identified.

It is rare in the developing world to find good water records available at block level, let alone at farm level. Thus, it is most likely that checks will need to be based on area. Cross-checks are preferably carried out during the course of the assessment, so that immediate correction may be made to the water charges. Some cross-checking will take longer, serving to evaluate the assessment performance in retrospect.

6.2 BILLING AND PAYMENT

The failure of government agencies to generate fee-based income may result from their failure to collect competently, thus compounding farmers' reluctance to pay.

The billing and collection exercises may be under-resourced. Typically, the accounts office is understaffed, staff are poorly paid, they lack transport, and are subject to widespread interference/intimidation to reduce or eliminate the charges.
Billing and charge collection should be treated as distinct activities, which need to be adequately planned and resourced. In setting up a billing and collection system, a number of critical choices need to be made:

- Timing of payment: before or after the delivery of irrigation supplies?
- Billing: delivering and receiving payment
- Payment: in cash only or also in kind?

6.2.1 Timing of billing

In much of South Asia and parts of China, all fees are collected at, or after, the end of the season, leading to considerable delays in payment. In practice, many smaller farmers do not have the cash to pre-pay for irrigation. This is probably a major factor behind the common practice e.g. in Bangladesh and Sudan, for farmers to pay in a share of the crop. In the Niger, the time limit for payment can be extended to six months after the end of the season, meaning that, even though actual fee collection rates are high, irrigation co-operatives are always short of operating capital and are vulnerable to real losses when inflation is high. When there are no effective enforcement mechanisms, the process of collecting dues can be time-consuming and frustrating. Section 6.3 demonstrates the cost of delay in payments in Macedonia.

If it is feasible, it is simplest to collect the charges before irrigation water is delivered. This system saves staff from having to pursue farmers for payment. Pre-payment for the service may be made prior to the irrigation season, as a fixed charge, or before each delivery.

Box 6.1 Pre-paid irrigation supplies

In the Chancay-Lambayeque system in North Peru, farmers pay a fixed price per '*riego*', a water delivery of 576 cubic metre. They receive a receipt against payment, which they show to the '*sectorista*' responsible for water deliveries in the secondary system. An automated administration of water payment is used, which links the payments to the scheduling of water turns. Since the introduction of the system of pre-payments, recovery of water charges has approached 100% in most years. In some years, recovery exceeded 100%, because the irrigation agency also sold water from the volume allowed for leakage and losses and also from savings where less than the 576 cubic metres per '*riego*' was supplied.

Source: Vos, 2002

Payment for each delivery is only workable when the system operates 'on demand'. A variation of the pre-payment method is used in Morocco, where farmers book and pay for a volume of water at the start of the irrigation season. In Jequetepeque, Peru, farmers pay per turn, but at the end of the season have to pay the balance if they have not used their full entitlement. The pre-paid delivery system only works when no water is transferred out of the system and water thefts can be kept in check. In the Chancay-Lambayeque system, the irrigation agency blocked a proposal for individuals at tertiary level to schedule water turns and collect water fees, fearing that it would not be able to control them. Instead, it perpetuated a system with "*sectoristas*" at secondary level who prepare water schedules and collect fees, which was less convenient for farmers who needed to travel further to book their turns.

Box 6.2 Pay-per-Irrigation, Turkey

To accommodate the farmers' ability to pay, irrigators could pay their water charges by instalments. The first quota could be paid for operation works at the beginning of the irrigation season, while the second quota, for maintenance and repair works, would be paid after harvesting when maintenance and repair work starts. The Irrigation Association of the Salihli Right Bank irrigation scheme, Turkey, implemented a pay-per-irrigation policy. It requires that before each of the first two irrigation turns, farmers must make an advance payment to the bank account of the Association and show the receipt to the head of the village irrigation committee before entering on the daily list of entitled irrigators. Once a farmer has paid for two turns, subsequent payments are on a credit basis with settlement later in the year.

6.2.2 Billing and payment

Effective presentation/delivery of bills is a crucial operation. If the bill is delivered at the time of assessment there is real scope for under-reckoning. Central processing allows cross-checks, but can be cumbersome in areas where there is no reliable postal system. Bills need to be delivered promptly, because if they are delivered long after the harvest season, farmers may be short of cash. In principle, it is reasonable to allow water users to pay their bills in a number of instalments, although the costs of administration increase, and where irrigation charges are low, it makes no sense to complicate payment in this way.

Box 6.3 Billing Procedures, Turkey

In Turkish government-managed irrigation systems, water charges are published two months before the irrigation season starts, so that the farmers are well informed. The individual bills are published in the village coffee houses and they are passed on to the irrigators, who then have two months to settle disputes, and make adjustments. Only then is the final billing established. Payments are made in two instalments.
However, if the job of collecting water charges and fines is made the responsibility of special collectors, and not of the irrigation provider, there is a need for close co-operation between these two units, and an incentive system for the collectors.

Source: Scheumann, 1997.

Under various existing systems, payment may be made to: field staff of the irrigation provider, the tax department or to special collectors; to the offices of the irrigation provider; to the tax department, or to designated banks. Special collectors, like the '*numberdar*' in Punjab, Pakistan, who is commonly a local notable, collect dues for a fee linked to collection (5% in Punjab).

Payment to banks usually involves a service charge, but is preferable to other modes of payment because it allows systematic overview of payments and is less open to fraud. However, a network of rural banks is required. Although it would seem desirable to allow various alternative modes of payment, in practice it may hamper oversight of the collection. The KBJNL Irrigation State Corporation in Karnataka, India, operates three different modes for the collection of water rates. Payment can be made at the agency sub-divisional cash counter, at designated banks, or directly to the concerned section officer of the Irrigation Department. As a result it is difficult to maintain oversight of payments. The overall collection record of the State Corporation has been poor.

Increasingly, cash payment is the norm, but in some systems in Eastern Europe and the former Soviet Union, as well as in countries such as Nepal, Bangladesh and Vietnam, payment in kind persists, although things are changing. Payment in kind is either the equivalent of a fixed charge (if the crop contribution is fixed), or is related to output (a proportion of production).

The drawback of payment in kind is that it requires a whole system of administration to handle, putting the burden of grading, storing and selling the crop output, with the associated risks, on the irrigation provider. It can be successful if there is a strong co-operative to handle the produce, as in Southeast Asia.

Increasingly, water user associations and private pump owners demand water payments in cash to meet their obligations in an increasingly cash-orientated local economy.

6.3 ACHIEVING GOOD REVENUE PERFORMANCE

Good revenue performance is reflected in accuracy of assessment and effective collection of charges.

Contrary to common belief, there is no straightforward relationship between collection rate and charge. In particular, there is no indication that a high charge necessarily makes recovery more difficult, provided that the charge is not suddenly, and dramatically, increased.

In some countries and systems, collections are close to 100%, for example, in China, Colombia, Mexico, Niger and Peru. However, the rates of recovery in many other places, including Albania, Bulgaria, India, Kazakhstan, Kosovo, Macedonia and Sudan can range between zero and 50%.

Contrary to common assumption, there does not appear to be a straightforward relationship between collection rate and charge. In particular, there is no indication that a high price necessarily makes recovery more difficult, provided that the price is not suddenly, and dramatically, increased. On the other hand, recovery is often low when charges are low, perhaps because there is little incentive to set up an effective collection system.

Under-assessment can be an important cause of poor recovery. In Sindh, Pakistan, for instance, collection rates used to be relatively high (85%); underassessment, by misclassification of cropped areas and by rebates for supposed disasters, was estimated to be of the order of 27 %. Similarly, at one stage in Albania, water was only supposed to be supplied after prepayment. In practice, water masters were liberal in providing water to those who approached them. The area under irrigation was several times larger than reported and most supplies were unaccounted for. Different water charging systems are more or less vulnerable to underassessment.

Poor rates of collection have serious implications, particularly for self-financing irrigation agencies which need a steady flow of income to stay in business. When water users in Surinam stopped paying water charges, water boards practically disappeared. By 1989 in India, fees collected on public systems in some states no longer covered the cost of collection. Poor irrigation supply performance begets poor revenue performance, as people refuse to pay.

6.3.1 Reasonable transaction costs

An irrigation charging system must operate satisfactorily at reasonable cost. Transaction costs - the cost of running the entire charging system - should be proportionate to the expected revenues. Though there is no generally accepted figure, transaction costs in excess of 20-25% of total charges appear high. When irrigation service fees were introduced in Indonesia in the early 1990s, it was expected that administrative costs and incentives would not exceed 20% of the gross collection. In practice, cumbersome administrative procedures led to greater costs. In these circumstances, the feasibility of the whole collection system needs to be radically reviewed. In Croatia, no charges are collected because the irrigated area and income are too small to justify the costs of the collection mechanism.

Box 6.4 Third party collection of water charges to reduce transaction costs

Concerns with the level of transaction costs have caused some countries to rely on existing tax collection mechanisms in order to collect water charges. This is particularly common in irrigation systems where no automated billing system exists. For example, in Egypt, the cost contribution for drainage investments is collected as a surcharge to the land tax at almost zero transaction cost.

Source: van Achthoven et al, 2003

A general overview of the transaction costs associated with implementing a water charging system is given Table 6.1 below. Not all categories of costs are relevant to all types of charging systems, but the summary may help to identify those which are relevant.

Table 6.1 Transaction costs in irrigation charging systems

Introduction costs	Running costs
Setting up finance department	Issue of bills and reminders
Training staff	In-service training
Initial awareness-building among stakeholders	Workshops, meetings, announcements
Developing new procedures, forms, records and standard contracts	Regular issue of forms, record sheets, contracts. Costs of reporting
Developing data base	Data entry. Updating registration records
Establishing asset register	Regular assessments
Purchase of basic facilities for field staff (transport, accommodation)	Transport and salaries of staff
Calibration and installing measuring equipment	O&M of measuring equipment
Intermediate organizations for bulk delivery (water users associations)	Transport and salaries of staff
Developing billing system	Monitoring and supervision of assessment and collection, internal auditing
	Bank charges on accounts
Establishment of regulatory system	Costs of regulatory system

6.3.2 Improving transparency in budgeting and spending

Transparency can be promoted by:

- An open budget, allowing scrutiny by interested users
- Review by audit panel
- Budgetary approval by a supervisory board or a regulatory committee.

In user-managed systems, supervisory boards and published budgets are sometimes adopted, though not yet universally. In agency-managed systems, such mechanisms are rare. It is particularly important that water users have confidence in the expenditures if large increases in charges need to be made.

Privatisation policies in the water supply sector have improved focus more generally on regulatory mechanisms, which are intended to safeguard the general interest in private or public supply of a monopoly good or service. Regulation may cover pricing, service performance, effects on the environment or other sectors. There is little privatization in irrigation, principally because schemes are rarely financially viable, and there are very few examples of regulatory bodies. Under the Sindh Water Management Ordinance 2002 in Pakistan, a regulatory authority was foreseen, but so far has not been put into effect (Box 6.5).

Box 6.5 Tasks for a regulatory authority in irrigation, Sindh province, Pakistan

1.General regulation
Ensure adequacy of regulations in the Sindh Irrigation and Drainage Authority (SIDA), the Area Water Boards (AWB) and Farmers' Organizations (FO) concerning, inter alia, elections, fines, penalties, cut-off of water supply.
Establish long-term and provisional regulations.

2.Financial regulation:
Ensure adequate financial management
Approve business plans of SIDA and AWB and annual financial statements/ audited statements – where the sustainability of the organization may be affected. Fix maximum charges

3.Registration of farmers' organizations
Register FOs and issue certificates of registration

4.Arbitration and trouble-shooting
Establish tribunals
Appoint acting Chairman
Receive and safeguard assets

5.Ensure adequate water management
Intervene in water distribution during drought
Intervene to avoid wastage of water

6.Communication
Identify problems of recurrent nature
Keep records for ready scrutiny

Source: Government of Sindh, 2002

Another example of a regulatory body comes from the introduction of irrigation service fees in Indonesia. At district level, the charging system was implemented by a Consultative Body, chaired by the district head or mayor, seconded by the head of the district planning board and the head of the irrigation agency, with representatives of water user federations. The Consultative Bodies - supported by a secretariat - decided on the O&M work program and also determined irrigation service levels and tariffs. Despite initial successes, the whole charging system subsequently failed, as it was complex to introduce and lacked political support.

As an alternative to institutionalised regulation mechanisms, operational and financial performance may be subjected to regular scrutiny by outside experts. Though not within the irrigation sector, the Water Conservation Mission established by the Government of Andhra Pradesh, India, to control water-harvesting programmes, introduced regular oversight by an expert panel.

It is recommended that regulatory mechanisms take strong account of national/regional practice and tradition rather than merely import practices from elsewhere.

Box 6.6 **Improved administration**

The introduction of a digital administration has helped improve transparency in water charging. In the Alto Rio Lerma District, Mexico, most WUAs ask for an annual external financial audit. They have started to use computers, which dramatically assist in the administrative and financial management of the associations. It is now easier to produce financial statements showing all income and expenditure. Similarly, when a farmer comes to pay his/her fees, it is very clear how many irrigation turns he has already paid for, how many he is still entitled to, for how many hectares and for what crops, and thus what are his total dues. Although financial malpractices have not disappeared, they appear to have decreased.

Source: Kloezen, 2002.

6.3.3 Improving transparency in service delivery

A number of mechanisms can help to promote transparency and accountability in the delivery of irrigation services:

- Benchmarking. Benchmarking is intended to allow the performance of irrigation providers to be assessed against internal or external norms and standards. In principle, if the underlying factors that can affect scheme performance are common, the relative performance of different schemes can be compared. FAO and World Bank have developed sets of possible benchmarks, from which a selection of those relevant to charging is included in Table 6.2. For the benchmarking process to become more than just the collection of performance indicators, the organizations must be committed by senior management to improved performance. Benchmarks should preferably be developed by the organization itself. The process of benchmarking is new in irrigation.
- Contracting-out of services helps promote transparency. In general, external parties are contracted, but contracts may also be drawn up between an irrigation provider and customers, also potentially between different units within an irrigation provider.
- Formal complaints handling procedures also promote transparency. A clearly-defined process will help reduce discontent and allow irrigation providers to identify actions vital to improving performance. Without formal complaints mechanisms, the reputation of the irrigation provider may suffer. It is preferable that a neutral party investigates complaints. The Ministry of Water Resources and Irrigation in Egypt is one of few irrigation providers with formal complaints handling procedures. The Ministry has a pre-printed format for farmers to register their complaints. Potentially, a systematic review of the complaints would also help in identifying priority works in canal or drainage system improvement, but this process is not operational as yet.
- Compensation, either for non-delivery of services or for damage by spills is another, more drastic, measure to promote transparency and accountability. Compensation arrangements have not been implemented in many places. They have been considered in some countries as part of service agreements between irrigation providers and water users. The compensation proposed for non-delivery of agreed supply is usually a discount on the water charges. Given the precarious financial position of most irrigation providers, this is the most that could reasonably be offered. In Macedonia, and several other countries, compensation is given for damage caused by leakage from canals - costing up to 5-10% of the total budgets.

Table 6.2 Indicators of financial performance

Indicator	Definition	Notes
Cost recovery ratio	Gross revenue collected / Total management + O&M cost (MOM)	Gross revenue collected from payments for water services by users. Total MOM costs of providing the services, excluding capital expenditure and depreciation/ renewals.
Maintenance cost to revenue ratio	Maintenance cost / Gross revenue collected	The cost of maintenance required to keep infrastructure in good order without build-up of deferred maintenance
Total MOM cost per unit area	Total MOM cost / Total serviced area	Total serviced area: the plan area actually served in the year.
Revenue collection performance	Gross revenue collected / Gross revenue invoiced	Gross revenue invoiced: total revenue due for collection from water users for services supplied.
Staff numbers per unit area	Total persons providing service / Total serviced area	Total number of personnel engaged in providing the services
Total MOM cost per unit volume of water delivered	Total MOM cost / Total annual volume delivered	Total annual volume delivered/year to water users, measured at a defined level of the system.

6.3.4 Improving collection

It may not be the irrigation charge as such, but farming fortunes in a given year - bad markets, failed harvests - that affect willingness to pay. In other cases, it may not the absolute price but a steep rise in price that creates resentment. Before introducing sanctions to enforce payment, the provider should aim to remove unsatisfactory aspects of service which discourage payment, such as poor water delivery and flawed billing processes. Other factors positively affecting users' willingness to pay include:

- Reasonable price
- Confidence that funds are well spent
- Effective enforcement of payments

Discounts for early payment are rare but could be considered (Section 5.2.2).

Particularly in places where there has traditionally been a lax attitude to charging and collection, water users will inevitably be suspicious of, and hostile to, the introduction of higher charges. Experience from other parallel programmes of change in irrigation, particularly turnover of system management to water users associations, strongly indicates that the irrigation provider or the revenue collector will need to set aside time and resources to explain, "sell" and introduce the increased charges. Field staff with skills in working with farmers are of great value.

It is suggested that the process of changing minds should be undertaken on a pilot basis, proceeding logically to:

- Identify basis for resistance
- Isolate genuine concerns
- Conduct stakeholder meetings to explain needs and identify constructive responses.
- Communicate issues more widely

Corruption is a risk in any institution handling large sums of money. The practice of paying bribes to secure irrigation supplies, or to get a reduction in the water bill, is well documented. Transparency and accountability must be improved in:

- Budgeting and spending
- Delivery of services
- Payment of water charges.

6.4 SANCTIONS FOR NON-PAYMENT

Unpaid bills are not necessarily the result of water users' reluctance to pay. They may also result from incorrect billing, failure to deliver bills or other mistakes by the revenue administration.

In many cases, however, non-payment is rife for want of effective sanctions.

Box 6.7 The cost of default

In Macedonia up to the year 1993, all defaulters were taken to court. However, when non-payment of dues became rampant in following years, court fees became so high that the financial position of the irrigation agencies was threatened. Recently, only groups of larger defaulters have been taken to court - without much success. Cases have to be initiated within three years of non-payment. There is a long delay (3-5 years) in the processing of court cases. By the time a decision is taken, inflation has taken its toll on the value of the dues. Many defaulters still refuse to pay – even when summonsed by the court. Unfortunately, there are few effective sanctions remaining unless the defaulter has a permanent source of income, which can be seized. As an example of the difficulties involved in going to court, one irrigation agency (Kocani) spent US $ 200,000 in court fees in 2000 and was able to retrieve only US $ 250,000.

Source: Hatzius, 2000

There are three main categories of sanction:

- penalties
- suspension of water deliveries
- legal action

The investigations needed to set up a charging system should establish that the judicial system can effectively support revenue collection, through the provision of legal sanctions in the case of non-payment. If it cannot, then the financial viability of the charging effort must be in doubt.

Penalties for non-payment or late-payment are relatively uncommon. Non-payment often persists and not is discouraged by fines. Fines generally work only when they are combined with legal action, which is only meaningful if the legal system is reasonably effective. The investigations needed to set up a charging system should establish that the judicial system can effectively support revenue collection. If it cannot, then the financial viability of the charging effort must be in doubt.

Box 6.8 Graduated sanctions enforced by Turkish Irrigation Associations

Conflicts among members and associations' management organs are mostly about damaging the structures, water distribution and payment of fees. Irrigation Associations are authorized to impose fines if farmers extract water out of turn or unlawfully. If members do not pay the fines or dues, graduated penalties are levied, e.g. first a warning, then a penalty, followed by the disconnection of services and a lawsuit. The nature of punishment depends on the seriousness of violations. For example, smaller farmers are not punished by some associations but merely warned for illegal water extraction. Some had problems with payment of bills from absentee landlords, who were uninformed about the existence and responsibilities of the Irrigation Associations. Then, they imposed penalties as well as realized that the absentees had to be informed. So notices of recovery were served and most defaulters paid. There is evidence that some associations have sent the default cases to court. However, so far none of them has refused service to its members.

Source: Scheumann and ul-Hassan. 2001.

Suspension of water deliveries is generally a more powerful sanction than financial penalties, and, moreover, to a greater or lesser degree lies within the control of the irrigation service provider.

Supply may be suspended at different stages of the cropping year but since payment is commonly sought only after the season, it is rare to suspend supply during a season. It is more common to withhold supplies at the start of the following season. Some irrigation providers create artificial scarcity by reducing deliveries to areas with a bad payment record.

Legal capacity to deny water to those who do not meet their obligations is essential for sustainable financing. In Argentina, revenue collection by water users associations only became effective after the WUAs were allowed to stop the delivery of water to defaulters.

However, if water users are billed individually, it is difficult to suspend supplies to individuals when they draw supply from a common watercourse. It is also effectively impossible on rice-based systems to suspend water to a single farmer, since water flows from field to field. Such cases ideally need to be dealt with by water user associations.

Legal sanctions should charge, or impose other penalty, for late payment. In Bangladesh and Pakistan, unpaid water charges are designated 'arrears of the land revenue' and are punishable by law. The Revenue Officer has magisterial powers and access to a wide range of sanctions – including imprisonment and confiscation of property. As a result, cost recovery is reasonably high - 95% in Sindh, for instance. In the USA and Australia, land has been seized as a last resort. Some irrigation agencies publish the names of defaulters, making it more difficult for them to acquire credit or to apply for public positions.

6.5 MONITORING AND FEEDBACK

The performance of the system set up to recover costs needs to be actively monitored on a regular basis to determine whether it is meeting established objectives. The results then need to be fed back to management and the audit panel (6.3.2), to assess whether policies are working and to identify possible corrective measures. The monitoring team should focus primarily on financial performance but will need to include technical personnel to ensure that charges made for, say, maintenance, are being spent correctly in priority areas.

Irrigation charging systems differ in the extent to which they allow both oversight and abuse. No system is foolproof and much depends on the culture, the control mechanisms and the will to tackle corruption and under-assessment.

As identified previously in the text, a number of practices reduce the risk of abuse, including:

1. Cross checking of assessments
2. Payment through banks rather than in cash
3. Payment in cash rather than in kind
4. Issue of receipts upon payment
5. Reduction of the number of assessment categories
6. Reduction of exemption categories and other special arrangements
7. Prompt action against defaulters.

6.6 CHAPTER SUMMARY

See Table 6.3

Table 6.3 Implementing an irrigation charging system – chapter summary

Step	Actions	Recommendations / notes
Assess the amount due	Overall basis for charging, and therefore the assessment process, should be determined according to policy objectives (See chapters 2 & 4).	Minimise the number of categories used, e.g. crop types, season, water source, provision of other services
	Identify users eligible for special provisions	This may be due to poverty, natural disaster or crop failure.
	Cross-check a sample of the assessments made	Ensure that the basis for assessment is clear and can be validated by both user and service provider.
Billing and payment	Consider individual pre-payment for a season or per irrigation. Alternatively, the service provider may bill a user group for an agreed volume(quota). This avoids problems of assessment for the service provider but the user group must determine how to charge individuals.	**Recommend** Pre-payment, using a 'fixed' parameter such as holding size. This avoids many of the problems associated with post season assessment, bill delivery and charge collection.
	Keep the three steps – assessment, billing, and receipt of payment – separate.	This may increase transaction costs but greatly improves transparency Do not issue the bill at the time of field assessment as this rules out any chance of checking the assessment.
	Use a single means of collecting payment: • Field agents • Through rural banks • At offices of service provider • At offices of user association	Where WUAs exist they should be the preferred means of collection. Otherwise rural banks, if an adequate network exists, potentially provide independent oversight. Always ensure payments are formally receipted. Ensure ease of access to point of payment.
Achieve good levels of recovery	Consider the value of an independent regulatory authority, potentially with powers to: • Formulate/advise on policy • Approve budgets and financial statements of service providers • Set and monitor standards of service • Arbitrate in disputes	
	Estimate the transaction costs (See Table 6.1)	Ensure that annual transaction costs are less than 20% of anticipated annual revenue.

Table 6.3 Implementing an irrigation charging system – chapter summary (continued)

	Establish transparent procedures to set budgets and permit stakeholders to review actual expenditures.	
	Establish procedures to measure service delivery, e.g. • Benchmarking • *Definition of service agreements* • Compensation for non-compliance with service contract • Formal complaints procedures	Most of these require a formal service agreement between the service provider and irrigators. These are not widely used but they *might greatly enhance users' willingness* to pay by providing a clear statement of what level of service is being paying for. Defining a quota assists in this (See 4.1.5).
	Invest resources to justify charging to users and explain the procedures.	Be prepared to modify procedures in the light of user concerns.
Specify sanctions for non-payment	Potential sanctions are: • Penalties – fines, public naming of defaulters • Suspension of supply • Legal action	Fines may only be effective when back by legal code. Effectiveness of public naming will depend on culture and public attitude towards service provider. Suspension of supply to individual users is difficult to impose. Supplying to and billing of user groups has advantage that suspending supply to a defaulting group may be technically feasible.

7. References:

Bosworth B., Cornish G., Perry C. and van Steenbergen F. 2002. *Water charging in irrigated agriculture: lessons from the literature.* OD145. December 2002. HR Wallingford Ltd. UK.

Burt C M. 2002. *Volumetric water pricing.* Draft document prepared for the Irrigation Institutions Window of the World Bank.

Cornish G. A. and Perry C. J. 2003. *Water charging in irrigated agriculture: lessons from the field.* OD146.November 2003. HR Wallingford Ltd. UK

Cornish, G., Bosworth, B., Perry, C. and Burke, J. 2004. *Water charging in irrigated agriculture: lessons from international experience.* Water Report 28. Rome: FAO.

Euroconsult 1998. Nara Pilot area Water Board, Pilot Institutional Study. Lahore

Government of Sindh 2002. Sindh Water Management Ordinance.

Hatzius T. 2000. *The Case of New Water Fee System in the Republic of Macedonia. Market and Non-Market Failure in Path Dependent Institutional Reform.* Division 45 – Rural Development. MAINTAIN – Case study no.7. Deutsche Gesellschaft für technische Zusammenarbeit (GTZ), Eschborn. 30pp.

Johnson S H III. 2001. A contribution by Sam H Johnson III. Irrigation Management Turnover Conference – Theme 6 (Financing Irrigation) – Intervention 97. International Email Conference on IMT, sponsored by FAO, INPIM and the Ford Foundation.
http://www.fao.org/landandwater/aglw/waterinstitutions

Kloezen, W. 2002. *Accounting for water: institutional viability and impacts of market-oriented irrigation interventions in Central Mexico.* PhD Thesis. Wageningen: Wageningen University.

Langford, K.J, Forster C L and .Malcolm D M. 1997. *Towards a financially sustainable irrigation system: lessons from the State of Victoria, Australia, 1984-1994.* World Bank Technical Paper 413. Washington DC: World Bank.

Molden, D. 1997, *Accounting for Water Use and Productivity,* SWIM paper No. 1, International Water Management Institute, Columbo, Sri Lanka

Scheumann, W. 1997. *Managing salinization: institutional analysis of public irrigation systems.* Berlin: Springer Verlag.

Scheumann W.and ul-Hassan. 2001. IMT: Irrigation Associations in the Lower Seyhan Plain and the Gediz River Basin (Turkey), e-conference FAO/INPIM.

Van Achthoven, A., Z. Merebet, K. Shalaby and F. van Steenbergen. 2003. *Agricultural Drainage: Toward an Interdisciplinary and Integrated Approach Egypt: A Case of Balancing Productivity and Environmental Pressure.* Washington: World Bank. www.worldbank.org/irrigation-drainage

Vos, J. 2002. *Metric matters: the performance and organisation of volumetric water control in large-scale irrigation in the North Coast of Peru.* PhD Thesis. Wageningen: Wageningen University.

Annexes

Annex A Applying the guidelines – a user checklist

The five tables of the checklist take the user through the logical series of steps set out in the body of the guidelines. The boxes on the checklist tables may not capture the full response to a question. For example, the user may not be able to describe in full the present situation relating to water availability and allocation for environmental/other demands in the small box provided. Rather, the checklist is intended to drive the user to gather sufficient information to be able to adequately describe the present situation. In going through that process it should become clear whether present policies and their outcomes at scheme level (Tables 1 and 2 respectively) are working well, and therefore need no change, or where policy is lacking or ineffectual and therefore changes are needed. The checklist tables provide a quick visual summary of current practices and where changes are needed, but the actual nature of that change may need extensive debate and planning before it can be implemented.

A worked example, based on information obtained for the Haryana case study (See Cornish and Perry, 2003), is included after the blank tables. The example serves to demonstrate the type of information required and how the tables may be used – it is not intended to be a definitive statement on what is presently practised in that state.

- Table 1 Policy Environment
  This table focuses on current policies and responsibilities. Often, the existing situation may be satisfactory with no changes required. Where changes are proposed, the proposals will usually depend on analysis and decisions that cannot be finalised until information on scheme level priorities and effects (Table 2) have been evaluated.

Section 3 of the table reviews which agency or management groups have control of the financial and management resources needed for the operation, maintenance, rehabilitation or improvement of irrigation and drainage infrastructure. Clear definitions of responsibility are required because the financial details of irrigation service charges (who pays, how much, to whom, and for what) must be linked to the level of service and the service provider.

Sections 4 to 7 of the table should prompt review of the policies of how an existing ISC is implemented. Chapters 5 and 6 of the guidelines should help to inform any proposed change in these aspects.

When completed, the table will summarise the present situation and any proposed changes, together with their legal, financial and administrative implications, that must be addressed at the political/ administrative level before field-level changes can be implemented.

- Table 2 Setting objectives: current practices and problem identification

This second table requires information at the level of an individual irrigation scheme, reflecting the practical outcomes of existing national or regional polices as they are applied.

Section 1 of the table reviews current water management practice and Section 2 gathers information on the financial performance of the scheme.

By assigning priorities to any problems or constraints that are identified it should become clear if solutions can be achieved through the revision or introduction of an ISC or whether it is first necessary to change or establish governing policies. Second, by ranking the current problems faced regarding water management and financial performance at scheme level it becomes possible to identify the key objectives of a proposed ISC.

- Table 3 Defining the basis for charging

Table 3 sets out the current basis for charging, and the revised proposal – relating this to the objectives. Again it is stressed that the check box structure of the table should lead the user to identify how any proposed change relates to agreed objectives that should be defined in the light of Tables 1 and 2. Furthermore, the implications of a change in rates, means of charging, collection, or use of funds must be related back to decisions required at the policy level (Table 1). The check box cannot contain a full resume of these issues but only a brief note or tick when that issue has been adequately addressed or defined. Where issues are identified, a separate analysis may be prepared and referenced in the checkbox.

- Table 4 Quantifying the charges

This table assumes the primary objective of the ISC is to recovery some proportion of costs. Whilst a charging method may be used that promotes water saving or increased productivity the guidelines caution against attempting to set the water price, and therefore the charge, on the basis of its value.

Sections 1 and 2 of the table refer back to the wider policy environment, checking that policy specifies what type and fraction of costs are to be recovered and what provision is made for costs not recovered through irrigation service charges. Sections 3 to 7 prompt the user to examine the mechanisms used to calculate costs, ensure they are regularly reviewed and define what levels of variation should exist. In many schemes accounting procedures may need to be changed to permit the transparent definition of costs.

The first two columns require the user to define current practice and identify whether or not it is effective in meeting policy objectives. Where changes are required in several aspects of the definition of costs, and therefore required revenue, the third column requires the user to indicate the order in which changes should be brought about

- Table 5 Implementation

This table provides a checklist of the key sections in Chapter 6 of the full guidelines and it should be used in association with that chapter. As with the earlier tables, the issues for checking are reduced to a brief heading but effective completion of the table may require considerable research and planning to implement effective change.

Table 1 Policy environment

	Present situation		Comments / Explanation
1. Responsibility for water resources planning, development and management			*The purpose of this section is to identify primary responsibility for the sector, and the general policy on Irrigation Service Charges*
2. Cost recovery policy for irrigation			
3. Water availability and allocation:			
Inter-sectoral priorities in allocation			*This information summarises the policies governing allocation of water to irrigation and other uses, any criteria that will affect the level or structure of charges, and the rules defining the nature of the irrigation service.*
Social criteria in water allocation between sectors			
Environmental/other demands			
Irrigation-specific policies, e.g. food self sufficiency, crop promotion			
Criteria for allocating water among irrigation users			
4. Financial and legal responsibilities for:			
System operation			*This information indicates who is responsible to pay for various aspects of system operation. Where there are farmer organisations, or delivery is to a group, there will often be two sets of responsibilities – above the group level, and within that level.*
Routine maintenance of infrastructure			
Rehabilitation or replacement of assets			
System improvement or modernisation			
5. Allocation of costs between irrigation and non-irrigation beneficiaries (where relevant.)			*Describe procedures for deciding who pays costs of irrigation services.*
6. Responsibility for assessment of Irrigation Service Charges			*Describe procedures for actually billing beneficiaries for services and how revenues from ISC managed and distributed.*
7. Responsibility for collection of Irrigation Service Charges			
8. Allocation of revenue from ISC			

Table 2 **Setting objectives: current practices and problem identification**

	Present Situation	Priority Ranking
1. Water Availability and Distribution		
Specification of project water entitlement (a)		*The purpose is to identify and prioritise problems with current service.*
Specification of individual users' water entitlement		
Availability of water (Supply/ demand balance) (b)		
Equity of distribution		
Rules for distribution of water (time, turn quantity...)		
Adherence to rules		
Productivity of water		
2. Current Costs, Charges and Revenues (c)		
Annual cost of actual O&M		*The purpose is to clarify current financial sustainability and to prioritise problems.*
Estimated annual expenditure for replacement		
Total amount billed		
Total amount collected		
Cause of differences		

Notes
a) This does not refer to estimates used in project feasibility and design but whether or not a fixed volume or fraction of available storage is allocated to permit seasonal or annual planning.
b) The supply/demand balance will vary between years of high, average and low supply. The purpose here is to gauge whether or not equating supply with demand is a priority objective.
c) Use consistent units in this section to permit comparison.

Table 3 Defining the basis for charging

<table>
<tr><th></th><th>Present Situation</th><th>Proposed Changes</th><th>Relationship to Objectives</th><th>Policy/Legal Implications</th></tr>
<tr><td>Actual volume of water received</td><td></td><td></td><td></td><td rowspan="6">Any proposed change should reflect and respond to the priority problems identified in previous tables. It is important to identify conflicts between proposed changes and current policy/law.</td></tr>
<tr><td>Proxy for volume of water received (crop type/area)</td><td></td><td></td><td></td></tr>
<tr><td>Farm area</td><td></td><td></td><td></td></tr>
<tr><td>Crop value</td><td></td><td></td><td></td></tr>
<tr><td>Other</td><td></td><td></td><td></td></tr>
<tr><td>Combination</td><td></td><td></td><td></td></tr>
</table>

Table 4 Quantifying the charges

<table>
<tr><th></th><th>Well-defined, Effective, Ineffective, Absent?</th><th>Priority ranking</th></tr>
<tr><td>1. Policy defines what costs to recover?</td><td></td><td rowspan="7">The purpose is to clarify the current basis for setting charges identifying and prioritising where policy or procedures are ineffective or absent.</td></tr>
<tr><td>2. Mechanism to meet costs not recovered from irrigation service charge.</td><td></td></tr>
<tr><td>3. How are O&M costs calculated?</td><td></td></tr>
<tr><td>4. How are capital and replacement costs calculated?</td><td></td></tr>
<tr><td>5. Mechanism to review and update charges.</td><td></td></tr>
<tr><td>6. Procedures to define variable levels of charge – within schemes / between schemes/ between types of user.</td><td></td></tr>
<tr><td>7. Mechanisms for compensation / rebates.</td><td></td></tr>
</table>

Table 5 Implementation

	Present situation	Proposed changes	Priority ranking
1. Basis for assessment:			
Consistent with policy objectives			
Accurate, with potential for cross-checking			
Number of categories			
2. Method and timing of billing			
3. Method of payment and receipting			
4. Transparency / accountability:			
In budget setting and expenditure			
In service delivery			
5. Sanctions for non-payment			

Table 1 Policy Environment – Haryana State, India

<table>
<tr><th></th><th colspan="2">Present situation</th><th>Comments/Explanation</th></tr>
<tr><td>1. Responsibility for water resources planning, development and management</td><td colspan="2">State Water Resources Committee comprising representatives of all water using sectors.</td><td></td></tr>
<tr><td>2. Cost recovery policy for irrigation</td><td colspan="2">Full costs of O&M of irrigation, drainage and other water related facilities to be recovered from beneficiaries except as noted at 3, (social criteria) below.</td><td>Policy is clearly stated and charges have been set to meet policy.</td></tr>
<tr><td>3. Water availability and allocation:</td><td colspan="2"></td><td></td></tr>
<tr><td>Inter-sectoral priorities in allocation</td><td colspan="2">1) Domestic water supply
2) Industrial
3) Irrigation</td><td>Charges reflect priorities, being much higher for non-agricultural use.</td></tr>
<tr><td>Social criteria in water allocation between sectors</td><td colspan="2">None, except that supplies to less-developed areas in south of state are exempted from recovery of pumping costs.</td><td></td></tr>
<tr><td>Environmental/other demands</td><td colspan="2">International and interstate agreements govern the quality of river-borne effluent. No regulation of quantities.</td><td>Major environmental issue is groundwater which is falling where fresh and rising where saline.</td></tr>
<tr><td>Irrigation-specific policies, e.g. food self sufficiency, crop promotion</td><td colspan="2">None. Water is rationed and farmers are free to choose their cropping pattern.</td><td></td></tr>
<tr><td>Criteria for allocating water among irrigation users</td><td colspan="2">Strict proportionality of water supply to land holding. Water right is linked strictly to the land, not the land-owner (or tenant).</td><td>Trading of water is illegal, but practised locally, within watercourses. Trading between watercourses is hydraulically impossible.</td></tr>
<tr><td>4. Financial and legal responsibilities for:</td><td>Major infrastructure above watercourse (100-200ha)</td><td>Watercourse</td><td></td></tr>
<tr><td>System operation</td><td>Irrigation Department</td><td>Farmers</td><td rowspan="4">While there is complete clarity about operational responsibilities, the financial situation is less clear, especially as the new policies require recovery of costs which in turn requires allocation of costs to specific purposes. This in turn will require changes to accounting procedures.
No clarity about treatment of improvements to major facilities.</td></tr>
<tr><td>Routine maintenance of infrastructure</td><td>Irrigation Department</td><td>Farmers</td></tr>
<tr><td>Rehabilitation or replacement of assets</td><td>Irrigation Department</td><td>Irrigation Department; farmers share costs.</td></tr>
<tr><td>System improvement or modernisation</td><td>Irrigation Department</td><td>Irrigation Department; farmers share costs.</td></tr>
<tr><td>5. Allocation of costs between irrigation and non-irrigation beneficiaries (where relevant.)</td><td colspan="2">Policy decision by State Water Resources Committee, subject to recovery of total O&M costs.</td><td>See above: no ongoing linkage between costs incurred (by purpose) and revenues.</td></tr>
</table>

<table>
<tr><td>6. Responsibility for assessment of Irrigation Service Charges</td><td>Irrigation Department</td><td rowspan="3">The system is effective in ensuring high levels of recovery, but complex due to involvement of two agencies (though the latter collects directly from farmers in any case). Again no linkage between payments and expenditures as revenues go to general funds.</td></tr>
<tr><td>7. Responsibility for collection of Irrigation Service Charges</td><td>Revenue Department</td></tr>
<tr><td>8. Allocation of revenue from ISC</td><td>To State general funds.</td></tr>
</table>

General comments:

1. Haryana's irrigation system is generally well run and well maintained. Scarcity is dealt with by a transparent procedure for rotating supplies among areas, which can be checked. Allocation to the state of Haryana is governed by interstate agreements, administered by a basin authority – the Bakhra-Beas Management Board.

2. Recent decisions to recover costs of O&M from beneficiaries have led to close attention to the substantial quantities of water supplied through the irrigation system to non-irrigation users. An approximate balance between revenues and expenditures has been achieved by setting higher charges for the non-agricultural users, who in turn are given priority in times of scarcity.

Table 2 Setting objectives: current practices and problem identification – Haryana State, India

	Present Situation	Priority Ranking
1. Water Availability and Distribution		
Specification of project water entitlement	Water allocated in accordance with inter-state agreements administered by Bakhra-Beas Management Board.	
Specification of individual users' water entitlement	Based on published rules dividing available water equitably over project area. Actual quota depends on availability at dam.	
Availability of water (Supply/ demand balance)	Surface water very scarce; groundwater over-exploited where fresh.	1 (but not relevant to ISC as groundwater is private property)
Equity of distribution	Generally good.	
Rules for distribution of water (time, turn quantity...)	Clear and publicly known.	
Adherence to rules	Good; enforced by farmers.	
Productivity of water	Generally quite high.	
2. Current Costs, Charges and Revenues		
Annual cost of actual O&M	1996-2000 average for all sectors $18m/yr. $6m recovered from irrigation; $12m from other sectors.	3. Basis for allocation between sectors is unclear.
Estimated annual expenditure for replacement	Not known.	2. If system is to become independent and sustainable, clarity on these issues is essential.
Total amount billed	Approximately $18m	
Total amount collected	Approximately $18m	
Cause of differences	Recoveries vary from year to year: if water supplies are low, farmers delay payment – but on average, costs are recovered.	

Table 3 Defining the basis for charging

	Present Situation	Proposed Changes	Relationship to Objectives	Policy/Legal Implications
Actual volume of water received				
Proxy for volume of water received (crop type/area)	X			
Farm area				
Crop value				
Other				
Combination		Crop-based plus fixed rate per hectare	Provide stable income to agency responsible for maintenance, plus charge based on benefit received	Combined charge can be levied as at present, plus land revenue component, but requires legislation to allow funds to go to responsible agency.

Table 4 Quantifying the charges

	Well-defined, Effective, Ineffective, Absent?	Priority ranking
1. Policy defines what costs to recover?	Yes, but limited to O&M; historical costs have been waived.	
2. Mechanism to meet costs not recovered from irrigation service charge.	Policy in place and currently working – but no long-term linkage between expenditures and charges.	3
3. How are O&M costs calculated?	General accounts of Irrigation Department	
4. How are capital and replacement costs calculated?	Not done.	1
5. Mechanism to review and update charges.	Need to continue to recover actual O&M, clarity on treatment of replacement costs.	
6. Procedures to define variable levels of charge – within schemes / between schemes/ between types of user.	Not transparent	2
7. Mechanisms for compensation / rebates.	Yes – where crops fail, charges are waived.	

General comments:
Clarity of who pays what, and how this relates to expenditures, needs further attention. If the linkage between payment and expenditure is "closed", the beneficiaries will take an interest in more cost-effective maintenance.

Table 5 Implementation

	Present situation	**Proposed changes**	**Priority ranking**
1. Basis for assessment:			
Consistent with policy objectives	Yes	Add fixed component.	
Accurate, with potential for cross-checking	Reasonable – similarity of service to all farmers will highlight problems in billing.		
Number of categories	1		
2. Method and timing of billing	End of season		
3. Method of payment and receipting	Written accounts as part of Land Revenue		
4. Transparency / accountability:			
In budget setting and expenditure	No	New accounting procedures are being introduced	1
In service delivery	Yes		
5. Sanctions for non-payment	Yes – recovered as arrears of Land Revenue		

Annex B Defining terms and understanding theory

There is a large body of academic literature discussing the application of advanced economic theory to water resources management, but much of it is opaque to the non-specialist reader. A relatively accessible review of the material is provided by Johansson (2000) whilst Dinar (2000) presents a useful collection of papers in his book, 'The Political Economy of Water Pricing Reforms'. This annex does not attempt to summarise that large body of material. Rather, it reviews the basic economic realtionship between demand and price and links this to to the field observations made in the case studies that support these Guidelines.

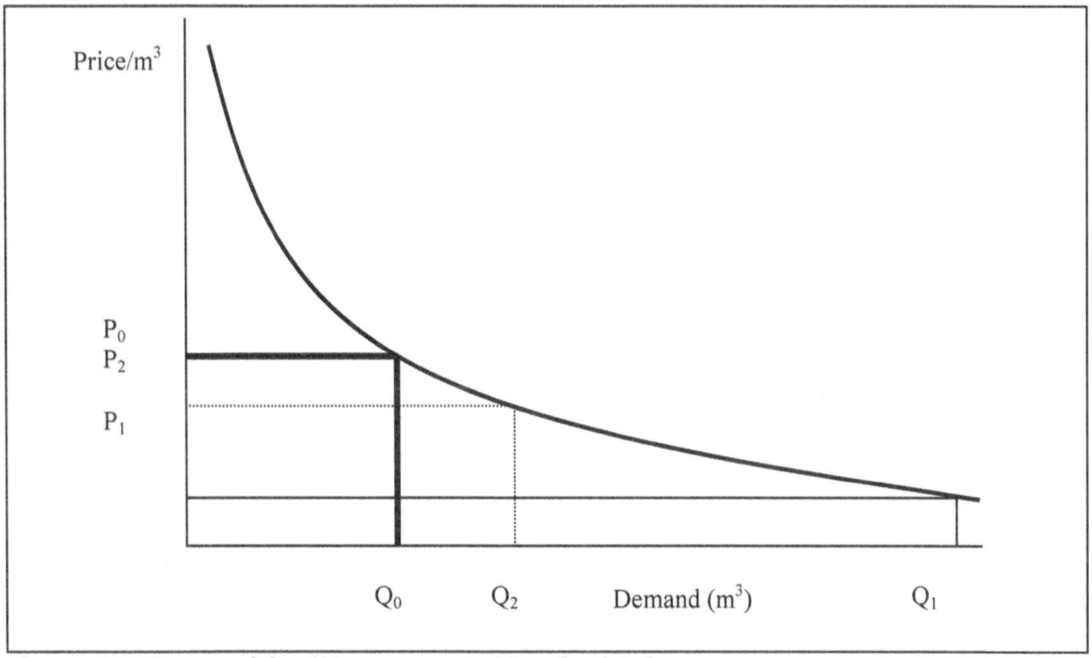

Figure A1: Demand for Water versus Marginal Price

Figure A1 shows the conventional relationship between price and demand. The shape of the demand curve is consistent with the idea that demand will fall with increasing price – we reduce consumption of something as it becomes more expensive (e.g. eat less apples), will look for alternative means of satisfying their requirements (e.g. buy oranges instead of apples) and, eventually will simply stop buying and using an input when its price exceeds the benefit derived from any use.

The concept of *marginal* price is central to this relationship between demand and supply: in the case of apples, the marginal price is the unit price of an additional apple. Frequently, irrigation charges are such that there is no precise relationship between number of units supplied and charge levied. Sometimes there is an approximate relationship; sometimes the payment is for "irrigation" without a clear definition of quantity. This is rather like paying for access to an apple orchard: once that price is paid, the entrant will take as many apples as he can eat (or carry). If the entrance fee is too high, demand will fall to zero; if it is too low, the trees will quickly be stripped of all the fruit. Markets with variable marginal unit prices are efficient ways to find the appropriate "market clearing" price, and as such are, at least theoretically, an attractive way to allocate scarce water resources.

If we assume that the sustainable quantity of water available is fixed at Q_0, then for any price *below* P_0 demand will *exceed* the available supply. If, on the other hand, the price of water is *higher* than P_o, then there will be more water available than is demanded by irrigators at that price. Starting from P_1 and Q_1, any increase in price will in theory reduce demand *towards* Q_0. The hypothetical example – represented by P_2 and Q_2 – shows a substantial fall in demand (about 50%) achieved through a substantial increase in price (about 200%). Nevertheless, demand remains substantially higher than the available supply, and if P_2 is the price set (perhaps on criteria such as cost recovery) additional measures would be needed to ensure that sustainability is achieved and consumption is reduced to Q_0. This point is critical: higher marginal prices will always reduce demand – but reducing demand to precisely the quantity of water that is available is a far more complex matter.

The precise relationships illustrated here are hypothetical – but they point to the central issue that will determine the impact of pricing on farmers' behaviour: How does the marginal price implied by an objective such as cost recovery (that is, required charge per hectare divided by the volume of water delivered per hectare) compare to the value of water[6]? If that price is still much lower than the value of water, which the case studies have demonstrated, then meeting a *cost recovery* objective may not do much to achieve the separate objective of reducing demand to a sustainable level, or encouraging more productive use.

The dominant considerations of managers and policymakers relate to financial sustainability, encouraging productive use, and balancing supply and demand for water. The first of these objectives is essentially an internal issue for the operating agency; the second and third are issues where the agency (and policy-makers more widely) want to influence behaviour, either through incentives, market forces or direct control.

They have a range of options to achieve these purposes, some of which can serve only one objective, others may serve more than one objective. At one extreme – serving only the cost recovery objective – is a basic irrigation service fee, unrelated to water use. At the other extreme is a rationing or quota system that fully ensures a balance between supply and demand – but implies no particular level of charging, and no assurance of financial sustainability.

Case studies (Cornish and Perry, 2003) illustrate these scenarios clearly[7]: in Macedonia, the irrigation charge is such that demand for water is lower than available supply; in Gujarat, the price of water is sufficiently high that farmers do not irrigate all their land – though they do continue to use more water than is sustainably available from the aquifer; and in each of the other case studies, demand exceeds supply at the marginal price, and is constrained by direct control of the water available.

The marginal price of water to irrigation users is often zero – this is true in all the case studies except for Gujarat and Morocco. The situation of a low marginal price is represented by P_1 and the associated demand by Q_1. Demand substantially exceeds the available supply – as in the case studies of Gujarat groundwater, in Haryana and Sindh (where the price of water is very low but physical rationing is enforced through the water management system), and Morocco (again, with the balance achieved through rationing). Where water allocation procedures are weak (as in Nepal), or poorly enforced (which happens in parts of Haryana) the insufficient supply is reflected in tail-end shortages.

The case of Haryana, where supplies are rationed but the marginal price is zero, illustrates the concept of *opportunity cost*. The farmers are constrained to Q_0 by the allocation procedures, and their behaviour is similar to what would be induced by a price of P_0: they manage water carefully, do not over-irrigate, and tend to grow crops with high returns to water. In Morocco, where the marginal price of water is relatively high, sustainable use is ensured through rationing, and again the farmers' behaviour (including use of sprinklers) is related to the opportunity cost of water rather than the actual price they have to pay for water.

---

[6] Parameters such as the price and value of water may seem abstract, but such indicators can often be calculated with reasonable accuracy from quite simple data. As an example, we take the case of Egypt: World Bank data tell us that GDP was \$98.5bn (2001) and agriculture accounted for 16.8% of GDP, or \$16.5bn. Water use in agriculture (FAO, Aquastat) (References)is about 54bcm, so that value added per unit of water consumed is approximately \$0.31/m^3. Payments for land and water charges in 2000 amounted to \$22M, so the equivalent price of water was only \$0.0004/m^3 .

[7] See Cornish and Perry (2003) for further information on the case studies referred to here.

Annex C Summary of the findings from the literature review and case studies that underpin these Guidelines

This annex summarises international knowledge and experience in charging for irrigation services, drawing from published literature and six commissioned case studies in five countries. Together, these sources provide a broad spectrum of theory and practice, from less-developed to more-developed countries. The full data and material which form the basis of this document can be found in two reports (Bosworth et al. 2002 and Cornish and Perry, 2003) which are outputs from a DFID-funded project "Irrigation charging, Water Saving and Sustainable Livelihoods". In analysing this material, the focus has been to identify the objectives that agencies set for their charging regime and to examine the extent to which different charging mechanisms have led to the realization of those objectives.

Policies of water pricing are affected by, and in turn affect, a large number of other important issues in the irrigated agriculture sector, for example, operation and maintenance; turnover and Water User Associations; rehabilitation and modernisation of systems; increasing competition for available water with other sectors/users; international trade and commodity pricing. Much attention has been devoted elsewhere to these matters. In contrast, although much theoretical work has been done on the economics of irrigation water pricing, there is still considerable lack of understanding of what impacts can be realistically expected from water pricing policies in practice, despite earlier publications. In order to focus attention on such a fundamentally important point, it was decided to confine the scope of this work to charging for defined objectives in irrigation, principally, for cost recovery and for limiting demand for water. Associated issues, including the ones set out above, were touched upon but were generally not dealt with in detail.

The purpose of undertaking new case studies, was to try to identify the realities of charging in practice, to obtain more secure basic data and to detect social, financial, institutional and technical factors which may constrain the effective implementation of pricing policies.

The main conclusions of the review and case studies are summarised here:
i) **Terminology**: The terms price, charge, value, cost, fee and revenue are widely and often interchangeably used in the literature. Often such terms are imprecise or open to more than one legitimate interpretation. In this review, "price" generally carries the implication of unit price - the actual or implied cost per cubic metre of water. Irrigation charges or fees relate to the overall payment that a beneficiary pays for the service - whether based on areas, volumes, crops, or whatever. Costs are always complex - "full" cost may imply some or all of: ongoing operation and maintenance; amortisation or recovery of capital costs; opportunity costs; social costs, and environmental costs.

ii) **The wide range of charges**: There are frequently large differences in charges and charging mechanisms within a single country, reflecting different objectives, different water sources, different degrees of water scarcity and irrigation schemes with different technologies, farm types or socio-economic objectives. Statements at a national level describing irrigation water charging must be regarded as indicative.

Price per cubic metre: There is a very large range in the reported volumetric price of water for irrigation. Prices as high as 18 to 29 US¢/ m^3, applied as a rising block tariff, are reported in Israel. Spain reports prices of 16 US¢/m^3 on schemes drawing from deep aquifers. In the market garden sector of Holland, where growers irrigate greenhouse crops from a municipal supply, the price per cubic metre may be as high US$1.30, but this is an extreme case. At the lowest end of the range Canada and Romania report prices below 0.1 US¢ / m^3. A price of about 2 US ¢ / m^3 (US$20 / '000$m^3$) is a common 'average' volumetric price charged for irrigation water, but these other values show the extent of the range.

Charge per hectare: Where irrigated area is used as the charging basis, comparison is made more difficult as it is not always clear in the literature whether figures quoted are seasonal or annual. Japan reports a figure of US$246 / ha; China and Greece report ranges of US$92–210 and US$50–150 respectively. US$40–50 /ha/year is a more representative 'average' charge in more developed countries. In India many states charge no more that US$10 /ha/year. Moreover, there is often considerable variation between theoretical or target rates and those actually charged in the field.

Collection efficiency: (Percentage of the billed amount that is collected.) Where information is provided, it indicates huge variation both within and between countries. For example, on the surface irrigation schemes of Bangladesh, collection rates are no more than 10 percent of the billed revenue, but on deep tube wells there is "almost full collection of revenues due". Of the countries where information on collection efficiency is reported, Mexico achieves the highest level with a national figure of 92 percent.

Proportion of costs recovered: There is more information available on this than on collection efficiency. The wealthier members of the OECD stand out as the few countries in the literature where there is full recovery of annual O&M costs and some recovery of capital costs. They include Japan, France, Australia, Spain and the Netherlands. However, in the overwhelming number of cases, water charging does not cover the annual O&M costs of irrigation schemes

iii) **Designing a charging system**: The objectives of a charging programme need to be articulated clearly in any discussion. The most widely pursued policy objectives are cost recovery and demand management. Macro-economic concerns of resource allocation between sectors, pollution charging and benefit taxation are recorded in the theoretical literature but they are seldom the drivers of national policies. Cost recovery and water demand management are two distinct objectives which require different types of intervention. However, it is surprisingly common to find substantial documents where these different objectives are apparently interchanged at random. This confusion, or blurring, of objective must be avoided so that policy makers, and those who advise them, have a clear understanding of what they are seeking to achieve and the tools that are relevant to that objective.

Where the objective is cost recovery, the range of costs that may be factored into the calculation is large. In practice, most agencies seek only to recover annual operation and maintenance costs. Non-volumetric water charges are simpler to administer than volumetric pricing as there is no requirement for extensive measurement infrastructure and continuous field recording. Volumetric water pricing or tradable water allocations (quotas) are used where the objective is to reduce or limit water use in the agriculture sector. However, there is little practical evidence from the field to support the view that volumetric pricing has a significant effect on farmers' water consumption patterns. Even in Jordan, Israel and Morocco, countries facing extreme water scarcity, the aim of water pricing is to recover service delivery costs. Volumetric water allocations, rather than water price, are used to ensure that demand is limited and other sectors' needs are met. In all of these countries, water is priced on a volumetric or approximate volumetric basis to indicate its value to users and discourage profligate use, but there is no attempt to use water pricing to achieve the balance between supply and the demand of competing sectors

The most widely used charging structure, which is adequate where the sole objective is cost recovery, is a fixed cost per hectare. In some cases, this may vary according to crop type, with higher charges for more water-demanding crops. Any price structure that contains a volumetric element is impractical where there is no infrastructure to routinely measure the volume used. Where this infrastructure does exist, a two-part tariff (with a fixed element to cover O&M costs and a variable element to reflect consumption) offers the benefit of assuring a more predictable basic income stream.

Water markets and tradable water rights could theoretically be more effective than water pricing as a means of achieving allocative efficiency. However, formal water markets may lead to inequitable access to water resources and disadvantage poor farmers, unless safeguards are provided to counter the tendency for water to flow according to purchasing power. Formal markets for large transactions between sectors require a well-defined legal and regulatory framework, as well as the infrastructure needed to move water from seller to buyer. They are found mainly in developed countries with Australia and Spain being widely cited examples.

It is concluded that recovery of O&M costs should generally not prove onerous to farmers, except for the poorest individuals and the poorest countries where special provisions/policies will need to be made. Nevertheless, farmers' dissatisfaction with levels of service and weak procedures for assessment, billing and enforcement commonly result in low levels of fee recovery. The principal constraint therefore appears to be in the management of systems and the administration of charging procedures in practice, rather than farmers' ability to pay.

iv) **The effects of charging on water saving**: The response in demand to volumetric water pricing is widely shown to be minimal. Current prices are well below the range where water saving is a significant financial consideration for the farmer. Volumetric prices may need to be 10-20 times the price needed for full supply cost recovery in order to significantly affect demand. It is also apparent that, whilst a number of countries use pricing to influence farmers' use of water below a defined ceiling, the ultimate control mechanism is by management of allocations, or quotas. Despite widespread use of price to control demand in the water supply sector, practical constraints have meant there are very few places in the world where price is the primary method of control in irrigation.

It is logical to suppose that farmers' responses are influenced by the relative magnitudes of the cost of water and its value to them. In some of the case study countries, the current cost of water is equivalent to a small percentage of their net crop income. However, in the Tadla scheme, Morocco, fees for surface water are some 15 percent of average net income, yet farmers will sometimes pay for additional and more expensive groundwater to supplement their quota. Therefore, it appears that water prices may need to be of the order of at least 20 percent of net income to begin to have significant impact on water use. In many countries, the rates currently paid are only a few percent of net income.

Even if it were feasible to supply water volumetrically, and to charge on an individual basis large numbers of small farmers growing cereals on Asian canal systems, there would remain the serious political and social difficulties of raising charges by something like an order of magnitude in order to begin to exert some measure of control.

As water becomes increasingly scarce, competition for water between the agriculture, municipal and industrial sectors will inevitably increase. Although the agriculture sector is seen as wasteful in its use of water, three important points must be made concerning these losses: (i) "lost" water often returns to an aquifer or river and can be accessed by other users. It is only "lost" if it deteriorates in quality or drains to a sink from which it cannot be economically recovered. Thus switching to 'high efficiency' irrigation methods such as drip or sprinkler may not result in significant overall savings of water if the previous losses were recaptured by others. (ii) Where excess withdrawals return to a river or an aquifer, the cost of service delivery is increased but overall levels of water scarcity may not be affected. (iii) The farmers' in-field management of water usually accounts for less than half of the 'losses'. As individual farmers have no control of the conveyance and distribution canals, pricing incentives do not affect these losses.

v) **Implementing charging polices**: Charging policies need to be formulated in full recognition of the various institutional and political factors which can limit cost recovery, including:

- The lack of political will to raise costs to farmers and slim down government agencies.
- The lack of motivation on the part of collection agencies, as fees return to the treasury and recovery is not linked to future funding.
- A vicious circle of low O&M expenditure leading to poor performance and increasing reluctance by farmers to pay.
- Insufficient resources for planning and implementing cost-effective charging mechanisms.
- Practical and political difficulties associated with enforcement of pricing policies.

The widely advocated policy of irrigation management transfer does not necessarily ensure recovery of full supply costs. The literature indicates that whilst turnover often leads to an increase in levels of cost recovery, revenues are still generally insufficient to cover full supply costs, as tariffs are set too low.

Where volumetric pricing is proposed to limit consumption, delivery must be measured and controlled to the individual user. In many developing countries, the service is provided to an aggregated group of farmers. Massive investments in re-engineering would be required to provide, even potentially, for "volumetric" delivery and pricing to each farmer. The challenge to administration and management would be unrealistic in the short to medium term.

The introduction of a water charging policy should be part of a larger package of measures designed to move to a virtuous circle where farmers are willing to pay for a good service, with the revenue being invested in sustained and improved service delivery. In the case of demand management, the literature again indicates that pricing is only a minor element. Legally recognized water rights and allocations and the use of tradable water rights are other common elements in such a programme.

IWA Publishing's authorised EU representative for General Product Safety Regulations is Diane D'Arras, 15 rue Duret, 75116 Paris, France, e-mail: safety@iwap.co.uk.

Printed and bound by CPI Group (UK) Ltd, Croydon, CR0 4YY
12/05/2026
02108893-0009